Hands-On Math

Learning Multiplication and Division
through Manipulative Activities

By
Dr. Kathleen Fletcher-Bacer

*Dedicated to my precious grandchildren: **Avalon** who at the age of 10 is quite the mathematician already; **Lincoln**, who is just learning to count at almost 2; **Lily** at 7 just discovering math; little **Riley** who gave me an opportunity to start writing while waiting for her birth, and to all my **future grandchildren** as I think there will be many more! Nana hopes your teachers will benefit from making math meaningful and fun!*

Order this book online at www.trafford.com
or email orders@trafford.com

Most Trafford titles are also available at major online book retailers.

Cover by Joseph Williams

Print information available on the last page.

ISBN: 978-1-4907-6761-1 (sc)
ISBN: 978-1-4907-6762-8 (e)

Library of Congress Control Number: 2016902880

Trafford rev. 03/07/2016

www.trafford.com
North America & international
toll-free: 1 888 232 4444 (USA & Canada)
fax: 812 355 4082

Table of Contents

About this book..vii

Basic Multiplication Facts
Using Physical Models for Conceptual Understanding
    Multiplying Counters 3
    Unlocking the Facts 7
    Linking Lego Array 10
    Multiplication Book 11
    Table It! 15
    Uncovering the Facts 18

Properties of Multiplication
Using Physical Models for Conceptual Understanding
    Closure Property 25
    Identity Property 27
    Zero Property 29
    Commutative & Associative Property 31

Number Theory for Multiplication
Using Physical Models for Conceptual Understanding
    Factor Hunting 37
    Are you my Multiple? 40
    Prime or Composite- That is the question!..................... 43

Basic Multiplication Concepts: Games & Activities
Developing Procedural Skills and Fluency
    What's my Fact?... 51
    Domino Time!.. 52
    Guess it? Keep 'em!.. 59
    Multi Matho 62
    Multiple Mascarades! 64
    Grapevine Time! 69
    Prime Time Detective 72

Multi-Digit Multiplication
Using Physical Models for Conceptual Understanding
    Graph it: 1x2 (one-digit number by a two-digit number)............. 77
    Graph it: 2x2 (two-digit number by a two-digit number)............ 81
    Extending Multiplication with Arrays: 3x3 (three-digit by two to
        three-digit number) 84
    Multi-Cheers! 88

Developing Procedural Skills and Fluency
    Managing Multi-digits...90
    Estimating Products..91
    Multi-Maze!...94

Basic Division Facts
Using Physical Models for Conceptual Understanding
    Carton It!...101
    Divide those Chips!..105
    Graph Paper Facts..109

Number Theory for Division
Using Physical Models for Conceptual Understanding
    Divisibility Rules for 2, 4, 5,8, 10
      How Does it End?...113
    Divisibility Rules for 3 & 9
      Sum Fun!...122
    Divisibility Rules for 6 & 7
      It's Complicated!...126

Basic Division Concepts: Games & Activities
Developing Procedural Skills and Fluency
    Sharing is Caring!..133
    Divide & Conquer!..134
    Detective Division..141
    AMAZEing Division 146
    Divisibility Rules!..151

Multi-Digit Division
Using Physical Models for Conceptual Understanding
    Division is for the Beans! [A] (two-digit by one-digit) 163
    Division is for the Beans! [B] (two and three-digit by
      two-digit numbers..170
    Extending Division with Group Arrays (two-to four-digit
      by two-digit numbers)..173
Developing Procedural Skills and Fluency
    Mastering and Managing Division with the Help of Monkeys! ...177
    Remainder? Run! (two-digit by one-digit) 181
    Headed Home!..184
    Tally Oh!...188
    Get Rich Quick!...190

Reinforcing Multiplication and Division
    Block It...199
    Multiply and Divide Rummy Style.................................204

Appendix A: **Manipulatives**
Manipulative Definitions and Resources 217

Appendix B: **Children's Literature that supports Multiplication and Division**
Multiplication.. 223
Division 226

ABOUT THIS BOOK

I hear and I forget.
I see and I remember.
I do and I understand.
Chinese Proverb

Hands-On Math is based on this proverb. The use of manipulatives is the key to enabling students to understand abstract mathematical concepts. The National Council of Teachers of Mathematics' Standards (NCTM) and the Common Core State Standards (CCSS) for Mathematics serves as a model for what should be occurring in the teaching/learning environments of mathematics and indicate that manipulatives are essential. Manipulatives can empower the NCTM (2000) standards and mathematical processes of Problem Solving, Reasoning and Proof, Communication, Connections, and Representation.

Learning how to effectively utilize manipulative activities in the classroom, however, can be very challenging. It is very important to use manipulatives in the manner they are designed – to develop those very important foundational mathematical concepts throughout the grade levels, thus making mathematics understandable to students. It has been my experience that children who discover foundational mathematical concepts for themselves retain those concepts and are able to transfer them beautifully to higher levels of mathematics. It is important to note, though, that if the manipulatives are not presented in the correct manner, they may actually confuse the student, rather than enhance learning.

The purpose of this book is to provide the teacher with a realistic approach to teaching multiplication and division for conceptual understanding with the utilization of manipulatives. The manipulative activities are designed to teach and develop the associated foundational skill and concept in conjunction with a textbook and/or other supportive materials.

Since multiplication and division skills are foundational for more difficult concepts within the Common Core State Standards (CCSS) Initiative (2014) of Operations & Algebraic Thinking, Number & Operations, Fractions, and the Number System, it becomes critical to build a solid foundation first.

This book is divided into specific skill categories and levels of mastery. The CCSS for mathematics are listed for grades 2-6 in each activity. Specific details of each of the aligned standards may be found at http://www.corestandards.org/Math
When you are in need of a lesson plan or a math activity for your students, simply choose one suited to the skill and standard you are teaching.

The symbol identifies the teacher pages. Each of these pages outlines a manipulative activity for you to follow. Quite often, a student page accompanies a teacher page so that the concept may be strengthened by student practice.

Each page labeled with a symbol invites students to investigate a concept. These student pages are best used initially with the teacher's direction. Once underway, however, student pages may be completed individually and checked for mastery by the teacher.

Appendix A includes a description of the manipulatives utilized in the activities and supportive information.

Appendix B is a collection of children's literature to support multiplication and division concepts.

INTRODUCTION

Multiplication was invented as a short method of adding equal addends. Instead of adding 5+5+5+5, students can learn that 5x4 yields the same result. Learning to use concrete and visual manipulatives for illustrating multiplication ideas is the first step in developing an understanding of this short method. The foundation for mastery of the full spectrum of multiplication becomes possible when students are able to construct the basic facts and understand the use of multiplication properties.

Similarly, division was invented as a short method for subtracting equal addends. All division problems can be done by subtraction, but as with multiplying by adding equal addends, the process is often long and tedious. Division is difficult to learn and to teach because, like subtraction, it is an inverse operation. In multiplication the factors are given and the product is unknown, but in division, the product and one factor are given; one factor is unknown. Building a solid foundation in multiplication facts becomes essential and transfers to division facts.

Once a solid foundation is established, students become fluent in multiplication and division basics to empower them with more complex problems that can transfer to algebraic thinking and higher level mathematics.

Get ready for math to become fun!

Basic Multiplication Facts

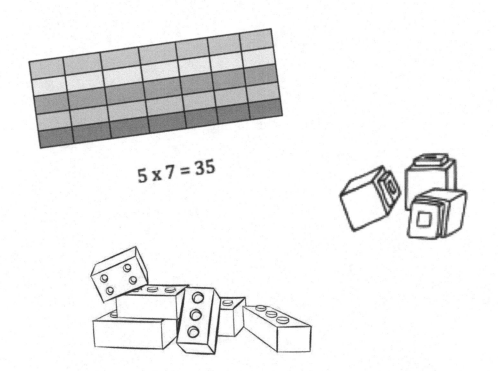

5 x 7 = 35

Building conceptual understanding of the basic multiplication facts is essential in helping students gain fluency with complex multiplication problems.

Basic Multiplication Facts:
Using Physical Models for Conceptual Understanding
Multiplying Counters

(**CCSS**: 2.0A.C.3; 3.0A.A.1)

Manipulatives Needed:
- *Counters* of any kind (e.g., buttons, beans, washers, beads, small blocks, etc...)
- Small bowls or containers for counters

Preparation:
- Distribute an ample supply of counters and up to 9 small bowls for each student or group of students.
- Copy and cut apart the *Multiplying Counters* cards (1 set for whole instruction and/or 1 set per group of students (pages 4-6)).

Procedure:
- Model a set of multiplication facts to the class. For example, model the 2 multiplication facts by illustrating with 2 bowls containing 1 counter each (2x1=2), 2 bowls with 2 counters (2x2=4) and continue through 2x9. As you model each fact, verbalize the mathematical sentence, e.g.,, I have 2 groups of 2 which equals 4.
- Shuffle the Multiplying Counter cards and use them to call out problems for students to represent with their counters and bowls. Students should share their findings.

Variation:
- Each group places a set of shuffled *Multiplying Counters* cards in the center of their workspace. Each student picks a card and creates the multiplication fact with their counters and bowls.
- Students should communicate their multiplication counter sets, e.g.,, I have 3 groups of 6, which equals 18.

Show me 3 x 6 (3 groups of 6). How many do you have altogether? 18

3x6

Multiplying Counters

[Card set to copy and cut out – page 1 of 3]

1x1	1x2	1x3	1x4	1x5	1x6
1x7	1x8	1x9	2x1	2x2	2x3
2x4	2x5	2x6	2x7	2x8	2x9
3x1	3x2	3x3	3x4	3x5	3x6
3x7	3x8	3x9	4x1	4x2	4x3

4x4	4x5	4x6	4x7	4x8	4x9
5x1	5x2	5x3	5x4	5x5	5x6
5x7	5x8	5x9	6x1	6x2	6x3
6x4	6x5	6x6	6x7	6x8	6x9
7x1	7x2	7x3	7x4	7x5	7x6

7×7	7×8	7×9	8×1	8×2	8×3
8×4	8×5	8×6	8×7	8×8	8×9
9×1	9×2	9×3	9×4	9×5	9×6
9×7	9×8	9×9			

Basic Multiplication Facts:
Using Physical Models for Conceptual Understanding

Unlocking the Facts

(CCSS: 2.0A.C.3; 3.0A.A.1)

Manipulatives Needed:
- *Unifix*™ *cubes* (up to 9 colors) or Interlocking cubes

Preparation:
- Supply each group of 4 students plenty of *Unifix*™ or Interlocking cubes. They will need up to 9 different colors to accommodate all the facts.
- Duplicate 9 copies of *Unlocking the Facts* (page 9) for each group to record their findings.

Procedure:
- Guide students through the process of constructing groups of colored cubes to represent the basic multiplication facts. It is essential to utilize the colors and the order of the fact, e.g., 3 groups of 5 in each group = 15. At this point do not introduce the communicative property. See the chart below to assist you in directing students to build various facts.

1's 1 group of 1,2,3,4,5,6,7,8,9 (Use 1 color)	**2's** 2 groups of 1,2,3,4,5,6,7,8,9 (Use 2 colors)	**3's** 3 groups of 1,2,3,4,5,6,7,8,9 (Use 3 colors)	**4's** 4 groups of 1,2,3,4,5,6,7,8,9 (Use 4 colors)	**5's** 5 groups of 1,2,3,4,5,6,7,8,9 (Use 5 colors)
1x1=1	2x1=2	3x1=3	4x1=4	5x1=5
1x2=1	2x2=4	3x2=6	4x2=8	5x2=10
1x3=3	2x3=6	3x3=9	4x3=12	5x3=15
1x4=4	2x4=8	3x4=12	4x4=16	5x4=20
1x5=5	2x5=10	3x5=15	4x5=20	5x5=25
1x6=6	2x6=12	3x6=18	4x6=24	5x6=30
1x7=7	2x7=14	3x7=21	4x7=28	5x7=35
1x8=8	2x8=16	3x8=24	4x8=32	5x8=40
1x9=9	2x9=18	3x9=27	4x9=36	5x9=45

6's	7's	8's	9's	
6 groups of 1,2,3,4,5,6,7,8, 9 (Use 6 colors)	7 groups of 1,2,3,4,5,6,7,8, 9 (Use 7 colors)	8 groups of 1,2,3,4,5,6,7,8, 9 (Use 8 colors)	9 groups of 1,2,3,4,5,6,7,8, 9 (Use 9 colors)	
6x1=6 6x2=12 6x3=18 6x4=24 6x5=30 6x6=36 6x7=42 6x8=48 6x9=54	7x1=7 7x2=14 7x3=21 7x4=28 7x5=35 7x6=42 7x7=49 7x8=56 7x9=63	8x1=8 8x2=16 8x3=24 8x4=32 8x5=40 8x6=48 8x7=56 8x8=64 8x9=72	9x1=9 9x2=18 9x3=27 9x4=36 9x5=45 9x6=54 9x7=63 9x8=72 9x9=81	

- Students record their findings on the *Unlocking the Facts* worksheets (page __)

$3 \times 5 = 15$ ⬅ red

⬅ blue

⬅ yellow

Special variations:
- You may also use this activity as a learning center in the classroom.
- If you don't have enough manipulatives, you can easily build the foundation through 5's and then have students use crayons to represent the rest of the facts using the Un*locking the Facts* worksheets. You may find this an effective strategy once they have conceptual understanding.

Answers: Use the fact charts to correct their work.

Unlocking the Facts

1's 2's 3's 4's 5's 6's 7's 8's 9's

1. Circle the multiplication fact above that you are working on.
2. Color in squares that match your work and write the multiplication sentence for each. For example, if working on 2's, use 2 colors and color in 2 groups of 1 each and write 2x1=2.

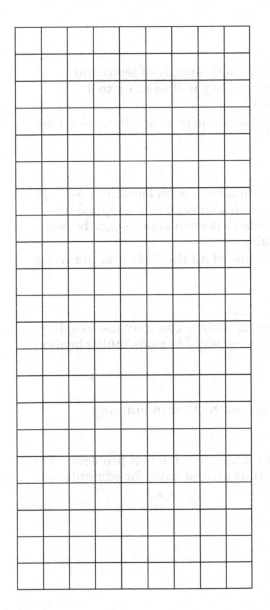

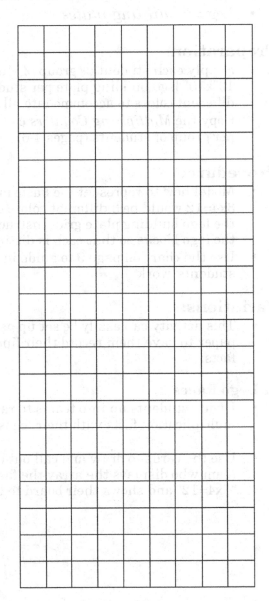

Name_____ Date_____

Basic Multiplication Facts:
Using Physical Models for Conceptual Understanding

Linking Lego™ Array

(CCSS: 2.0A.C.3; 2.0A.C.4; 3.0A.A.1)

Manipulatives Needed:
- *Lego™ blocks* (up to 9 different colors)
- *Lego™ building plates*

Preparation:
- Supply each student or group of students an ample supply of legos and 1 10"x10" lego building plate per student/group. They will need up to 9 different colors to accommodate all the facts.
- Copy the *Multiplying Counters* cards (1 set for whole instruction and/or 1 set per group of students (pages 4-6).

Procedure:
- Model how to represent the basic multiplication facts with legos. For example 3x4=12 would be 3 different colors of legos with 4 in each row snapped into the lego building plate grid. Instruct students to leave a small space between the lego blocks so that each is distinguishable.
- Use the chart on page 9 to guide in the building of all the facts and to correct students' work.

Variations:
- This activity can easily be set up as a learning center. You may use graph paper to have them record their findings as they build various multiplication facts.

Lego Races
- Divide students up into teams to race against each other in building multiplication facts with their legos.

- Use the cards to draw and call out different multiplication fact problems. The team who displays the array the fastest stands up and says, for example, "3x4=12" and shows their board to the class.

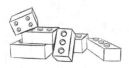

Basic Multiplication Facts:
Using physical models for Conceptual Understanding

Multiplication Book

(CCSS: 2.OA.C.3; 2.OA.C.4; 3.OA.A.1)

Manipulatives Needed:
- 1 in graph paper (10 sheets per student) (page 13)
- 1 cm graph paper (10 sheets per student) (page 14)
- 9 different colors of crayons, markers, or colored pencils
- 9x12 in. light-colored construction paper (15 sheets per student)
- Scissors
- Glue or glue sticks
- 9 different colors of 1" colored squares (for teacher demonstration)

Preparation:
- Supply each student with all the above materials except for the colored squares.

Special Note: I use different sizes of graph paper according to the grade level. Use the larger squared graph paper for younger students and the smaller squared graph paper for older students. You can also mix and match sizes so that they fit on the pages better. For example, use larger squared graph paper for 1's – 5's and smaller squared graph paper for 6 - 9 multiplication facts.

Procedure:
- Using the colored squares on 1" graph paper, illustrate how they will create the pages of their multiplication books. For example, in creating the 3's page, students will use 3 colors to fill in horizontal rows of designated amounts, e.g.,, 3 rows of 1,2,3,4,5,6,7,8,9. Instruct students to cut out 3x1, 3x2, 3x3, etc… and glue them onto on sheet of construction paper and then write the number sentence beside each representation of the multiplication fact.

r			3x1=3
g			
y			

r	r		3x2=6
g	g		
y	y		

r	r	r	3x3=9
g	g	g	
y	y	y	

(r=red, g=green, y=yellow to indicate colors as an example)

- Note that for each new subsequent multiplication set of facts, another color will be added, thus 4's will have 4 colors and so forth until the 9's with 9 colors.
- Students will create a page/s for each multiplication set of facts, thus they will have a page for 1's, 2's, 3's, 4's, 5's, 6's, 7's, 8's, and 9's.

- Monitor students' work carefully. Make sure they are showing each fact correctly. They will begin to see patterns and the properties of multiplication.
- After they have created all their pages representing the multiplication tables through 9, have them create a cover page titled, *"My Multiplication Book"* with their name and any design they want to add.
- Staple/bind all the pages together.

Special note: This book can now serve as a wonderful manipulative tool for not only learning the basic facts but also conceptual understanding of number theory concepts (,. factors, multiples, prime and composite). You will notice that students will begin to visualize the facts and patterns. Some will comment that they can see the basic multiplication facts in their heads!

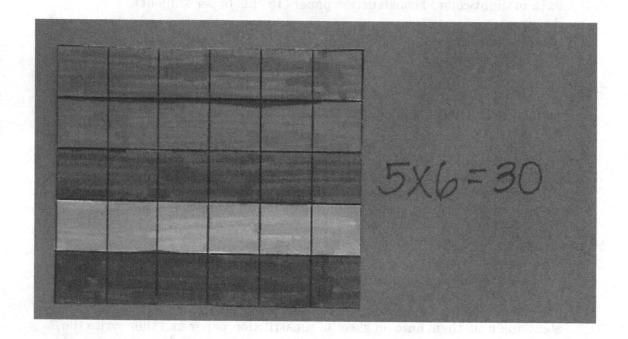

One-Inch Graph Paper

One-Centimeter Graph Paper

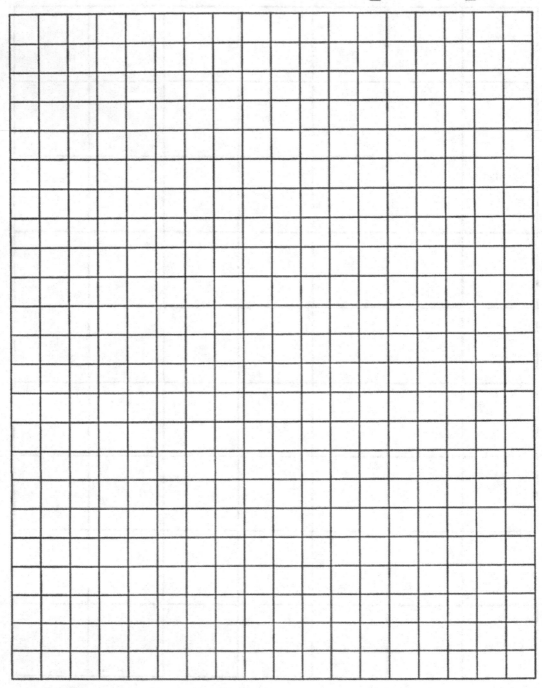

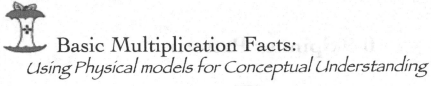

Basic Multiplication Facts:
Using Physical models for Conceptual Understanding

Table it!

(CCSS: 2.0A.C.3; 3.0A.A.1)

Manipulatives Needed:
- Two *0-9 Spinners* per student
- *Multiplication books* created from pages 11-12.
- *Table It* worksheet

Preparation:
- Assemble the spinners (page 16) and distribute two to each student.
- Provide each student with a copy of the *Table It!* worksheet (page 17).
- Student multiplication books can be used to reinforce multiplication facts.

Procedure:
- Instruct students to write numbers 0-9 in random order along the top and left side of the *Table it* worksheet.
- Students will spin each spinner to create their multiplication fact. After they find that fact in their books, write the answer in the correct square on their worksheet.
- Continue activity until they have completed the table.

Variations:
- Set a time limit and see who can create the largest number of different combinations and fill up their worksheet the fastest.
- Have students work in groups of 2-4 with one worksheet, each taking turns to spin and solve.

Special note: This method of utilizing the multiplication book and working with the multiplication facts out of order helps with concrete visualization of the basic facts. Answers will vary according to how each student set up their chart. Even though assessing their work will take a little longer, it is worth the effort.

8x6=48

0-9 Spinner Pattern

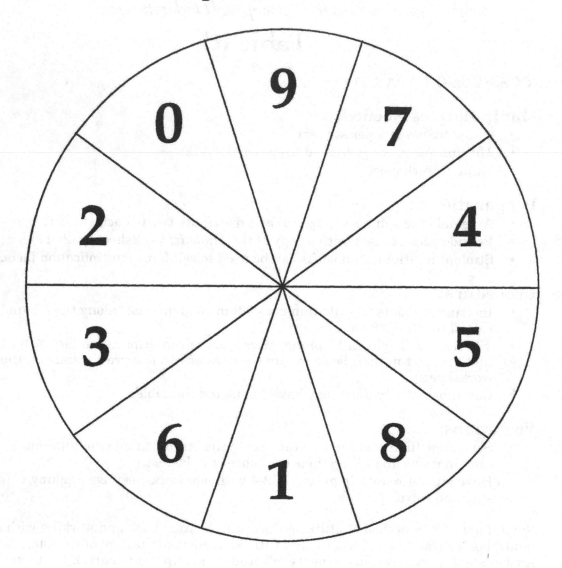

To assemble, duplicate the pattern onto cardstock. Use a brad to secure safety pin in center of spinner.

Table it!

X									

Name_____ Date_____

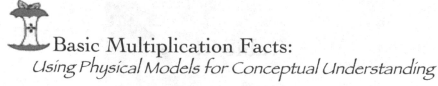

Basic Multiplication Facts:
Using Physical Models for Conceptual Understanding

Uncovering the Facts

(**CCSS**: 3.0A.A.1; 3.0A.A.3)

Preparation:
- Print the *Uncover the Fact Finder* pattern onto cardstock for each student and cut along the dotted lines (page 19).
- Provide each student with a copy of the *Uncover the Fact* worksheet

Procedure:
- Students place their *Uncover the Fact Finder* on top of the *Uncover the Fact Grid* worksheet, revealing any size rectangle. This rectangle represents a multiplication fact with factors (vertical and horizontal numbers on grid). The product is found by counting the squares.
- Students use the *Fact Finder* moved to different positions to reveal all the different multiplication facts and fill in the fact grid with products by counting the squares in each set of factors. (e.g.,, 3x7=21 shown below)

Uncover the Fact grid

x	1	2	3	4	5	6	7
1							
2							
3							

Uncover the Fact Finder

Answers:

x	1	2	3	4	5	6	7	8	9
1	1	2	3	4	5	6	7	8	9
2	2	4	6	8	10	12	14	16	18
3	3	6	9	12	15	18	21	24	27
4	4	8	12	16	20	24	28	32	36
5	5	10	15	20	25	30	35	40	45
6	6	12	18	24	30	36	42	48	54
7	7	14	21	28	35	42	49	56	63
8	8	16	24	32	40	48	56	64	72
9	9	18	27	36	45	54	63	72	81

Uncover the Fact Finder

Cut out this dotted rectangle

Uncover the Fact Grid

	1	2	3	4	5	6	7	8	9
1									
2									
3									
4									
5									
6									
7									
8									
9									

Name_____ Date_____

Properties of Multiplication

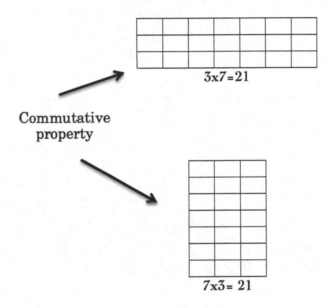

Commutative property

3x7=21

7x3= 21

Building conceptual understanding with the closure, identity, zero, commutative and associative properties of multiplication enables students to gain fluency with the application of basic multiplication facts.

Properties of Multiplication:
Using Physical Models for Conceptual Understanding

# Closure Property
[axb=c]

(**CCSS**: 3.0A.B.5; 3.0A.D.9)

Manipulatives Needed:
- *Counters* of any kind (e.g., buttons, beans, washers, beads, small blocks, etc...)
- Small bowls or containers for counters
- 1 in or 1 cm graph paper (see pages 13-14 for graph paper)
- Colored pencils, markers or crayons
- *Multiplication books* created from pages 11-12

Preparation:
- Distribute manipulatives to individuals or groups of students. You may also use the manipulatives for class demonstration.

Background concept – The closure property of multiplication states that axb=c (e.g., If you multiply any two whole numbers (excluding 1 or 0), you will get a new unique number).

Procedure:

Closure property with counters and containers:
- Guide students through the process of understanding the closure property by using counters and containers. For example, ask students to show you 3 containers (groups) with 2 counters in each. How many do we have? 6. 3x2=6. 5 containers with 3 counter in each. How many? 15. 5x3=15. Continue until students are able to deduct that any time 2 numbers (exclusive of 0 and 1) are multiplied together, a new unique number is created.

2x8=16   3x4=12

Closure property with graph paper and Multiplication books:

- Use graph paper to illustrate various rectangles representing multiplication facts exclusive of 0 and 1. (e.g.,, mark off 3 groups of 7 – 3x7. How many do we have? 21. 8 groups of 2 – 8x2. How many do we have? 16.

- An excellent way to illustrate the closure property is for students to use their multiplication books they created on page ___ and find all the multiplication sentences that are an example of the closure property.

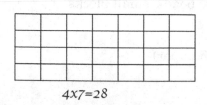

4x7=28

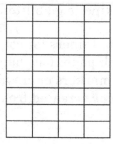

8x4=32

3x5=15

Properties of Multiplication:
Using Physical Models for Conceptual Understanding

# Identity Property
[ax1=a/1xa=a]

Teaching Tip for students: *You never lose your identity when you multiply by 1.* ☺

CCSS: 3.0A.B.5; 3.0A.D.9)

Manipulatives Needed:
- *Counters* of any kind (e.g., buttons, beans, washers, beads, small blocks, etc…)
- Small bowls or containers for counters
- 1 in or 1 cm graph paper (see pages 13-14 for graph paper)
- pencils, markers or crayons

Preparation:
- Distribute manipulatives to individuals or groups of students. You may also use the manipulatives for class demonstration.

Mathematical concept – The identity property of multiplication states that ax1=a or 1xa=a. (i.e., if you multiply any number by 1 you will always end up with that number).

Procedure:

Identity property with counters and containers:
- Guide students through the process of understanding the identity property with counters and containers. For example, ask students to show you 3 containers (groups) with 1 counter in each. How many do we have? 3. 3x1=3. 5 containers with 1 counter in each. How many? 5. 5x1=5. Show 1 container with 7 counters. How many counters? 7. 1x7=7. Continue until students are able to deduct that any time 1 is a part of our groups in multiplication, the value remains the number (of counters) we have.

1x8=8   3x1=3

Identity property with graph paper:

- Use the graph paper to illustrate various rectangles representing groups of 1. (e.g., mark off 1 group of 7 – 1x7. How many do we have? 7. 8 groups of 1 – 8x1. How many do we have? 8. Continue outlining various multiplication problems all using a factor of 1. Continue until students are able to understand the principle of the identity property.

Once students are able to apply this concept you can have lots of fun with multiplying large numbers by 1. 1x5,439 = ? 5,439! This is great boost for their attitudes towards math.

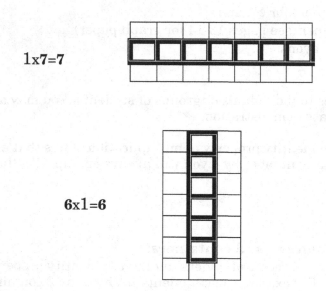

1x7=7

6x1=6

 Properties of Multiplication:
Using Physical Models for Conceptual Understanding

# Zero Property
[Ax0=0 or 0xa=0]

Teaching tip for students: *If you multiply any number by (zero) nothing, you will end up with nothing!* ☺

(**CCSS**: 3.0A.B.5; 3.0A.D.9)

Manipulatives Needed:
- *Counters* of any kind (e.g., buttons, beans, washers, beads, small blocks, etc...)
- Small bowls or containers for counters
- 1" or 1 cm graph paper (see pages 13-14 for graph paper)
- Colored pencils, markers or crayons

Preparation:
- Distribute manipulatives to individuals or groups of students. You may also use the manipulatives for class demonstration.

- *Mathematical concept* – The zero property of multiplication states that $nx0=0$ and $0xn=0$ (i.e. if you multiply any number by 0 you will always end up with 0)

Procedure:

Zero property with counters and containers:
- Guide students through the process of understanding the zero property with counters and containers. For example, ask students to show you 3 containers (groups) with 0 counters in each. How many do we have? 0. 3x0=0. 5 containers with 0 counters in each. How many? 0. 5x0=0. Show 1 container with 0 counters. How many counters? 0. 1x0=0. Continue until students are able to deduct that any time 0 is a part of our groups in multiplication, the answer (product) will always be zero.

1x0=0   3x0=0

Zero property with graph paper:

- Use the graph paper to illustrate what happens when you multiply any number by zero in trying to show the number of squares multiplied by 0. (e.g., mark off 3 groups of 0 – 3x0. How many do we have? 0. 8 groups of 0 – 8x0. How many do we have? 0. Continue outlining various multiplication problems all using a factor of 0. Continue until students are able to understand that it is impossible since the answer is always zero.

Once students are able to apply this concept you can have lots of fun with multiplying large numbers by 0 just like with 1. 0 x 45,439 = ? 0. This is another great boost for their attitudes towards math.

0x7=0

Properties of Multiplication:
Using Physical Models for Conceptual Understanding

Commutative & Associative Property
[a x b=b x a]

[(a x b)x c = a x (b x c)]

Teaching tip for students: *Order doesn't matter.* ☺

(**CCSS**: 3.0A.B.5; 3.0A.D.9)

Manipulatives Needed:
- 1 in or 1 cm graph paper (see pages 13-14 for graph paper)
- Colored pencils, markers or crayons
- scissors
- *Multiplication books* created from pages 11-12
- *Who's my Sidekick?* worksheet

Preparation:
- 2 copies of *Who's my sidekick?* worksheet for each student (page 31).
- Distribute manipulatives to individual or groups of students. You may also use the manipulatives for class demonstration.

Mathematical concept – The commutative property of multiplication states that axb=c or bxa=c (i.e. it doesn't matter in which order you multiply the numerals, your product will always be the same). Once students understand this concept they can easily transfer it to the associative property with more than 2 factors.

Discussion – Have students give examples of when order is and is not important for certain tasks (e.g., socks on before shoes is important; brushing teeth before hair is not, as the outcome will be the same). With commutative, order is not important. Let's explore this property.

Procedure:

Commutative property with graph paper:
- Guide students through the process of understanding the commutative property by having them illustrate various multiplication arrays on their graph paper. For example, ask students to illustrate 6x5 and 5x6. Cut out each array. Count the squares inside (30). Rotate one so it fits on top of the other to illustrate they both have the same product.
- Continue with various arrays until students see the pattern and they are able to generalize what happens to the product when we switch the order in which we multiply the same factors.

Commutative property with multiplication books:

- The multiplication books created on pages _____ are excellent for students visualizing all the facts that have, what I call, a *commutative sidekick* ☺. Instruct students to go through their books and find all the multiplication facts that have a *sidekick* and complete the *Who's my Sidekick* worksheet.

Multiplication fact	Sidekick	Multiplication fact	Sidekick
2x1=2	1x2=2	3x1=3	1x3=3
2x3=6	3x2=6	3x4=12	4x3=12
2x4=8	4x2=8	3x5=15	5x3=15
2x5=10	5x2=10	3x6=18	6x3=18
2x6=12	6x2=12	3x7=21	7x3=21
2x7=14	7x2=14	3x8=24	8x3=24
2x8=16	8x2=16	3x9=27	9x3=27
2x9=18	9x2=18		

- After they have listed all the facts and sidekicks, have them write the 8 remaining facts that do not have a sidekick, although these facts still illustrate the property. (i.e. 1x1, 2x2, 3x3, 4x4, 5x5, 6x6, 7x7, 8x8, 9x9)

Answers:

Fact	Sidekick	Fact	Sidekick	Fact	Sidekick
2x1=2	1x2=2	4x1=4	1x4=4	7x1=7	1x7=7
2x3=6	3x2=6	4x5=20	5x4=20	7x8=56	8x7=56
2x4=8	4x2=8	4x6=24	6x4=24	7x9=63	9x7=63
2x6=12	6x2=12	4x7=28	7x4=28	8x1=8	1x8=8
2x7=14	7x2=14	4x8=32	8x4=32	8x9=72	9x8=72
2x8=16	8x2=16	4x9=36	9x4=36		
2x9=18	9x2=18	5x1=5	1x5=5		
3x1=3	1x3=3	5x6=30	6x5=30		
3x2=6	2x3=6	5x7=35	7x5=35		
3x4=12	4x3=12	5x8=40	8x5=40		
3x5=15	5x3=15	5x9=45	9x5=45		
3x6=18	6x3=18	6x1=6	1x6=6		
3x7=21	7x3=21	6x7=42	7x6=42		
3x8=24	8x3=24	6x8=48	6x8=48		
3x9=27	9x3=27	6x9=54	9x6=54		

Who's My Commutative Sidekick?

Multiplication fact	Sidekick	Multiplication fact	Sidekick

Name_____ Date_____

Number Theory
For Multiplication

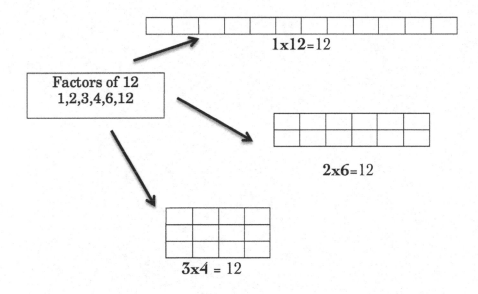

Number theory is essential in helping students understand the relationship and characteristics of numbers while providing an important foundation for operations and algebraic thinking.

.

Number Theory for Multiplication: Factors
Using Physical Models for Conceptual Understanding

Factor Hunting

Teaching tip for students: *Factors are the facts behind the answer (multiple / product).* ☺

(**CCSS**: 3.0A.A1; 4.0A.B.4; 6.NS.B.4)

Manipulatives Needed:
- *Multiplication books*
- *Factor Hunting* worksheet

Preparation:
- Duplicate up to 6 copies of the *Factor Hunting* worksheet for each student (page 36).
- Students will utilize their multiplication books created from pages 11-12 to hunt for factors.

- *Mathematical concept* – Factors refer to the numbers multiplied together to generate a new number (product/multiple). The easiest way for students to link the concept to the name is to think of them as the **facts** behind the answer.

Procedure:
- Instruct students to use their multiplication books, the clues you will provide, and the *Factor Hunting* worksheet to hunt for factors.
- Provide students with any or all of the following clues to hunt for factors:
    1. Find all the multiples where the factors are the same (e.g., 4x4=12).
    2. Find all the multiples with both factors less than ___ (specify a number).
    3. Find all the multiples with both factors greater than ___ (specify a number).
    4. Find all the multiples where both factors are even numbers.
    5. Find all the multiples where both factors are odd numbers.
    6. Find all the multiples where one factor is even and one factor is odd.

Clue - Multiples where both factors are odd

Factors	Product (answer or multiple)
3 x 1	3
3 x 3	9
3 x 5	15

Factor Hunting!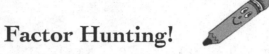

Write the clues given to you by your teacher in the chart below and hunt for factors.

Clue _____

Factors	Product (answer or multiple)

*Name*_____ *Date*_____

Factor Hunting **Answer sheet**

Clues:

1. Find all the multiples where the factors are the same.
   1x1, 2x2, 3x3, 4x4, 5x5, 6x6, 7x7, 8x8, 9x9

2. Find all the multiples with both factors less than ___ (specify a number).
   Answers will vary

3. Find all the multiples with both factors greater than ___ (specify a number)
   Answers will vary

4. Find all the multiples where both factors are even.
   2x2, 2x4, 2x6, 2x8, 4x4, 4x6, 4x8, 6x6, 6x8, 8x8

5. Find all the multiples where both factors are odd.
   1x1, 1x3, 1x5, 1x7, 1x9, 3x3, 3x5, 3x7, 3x9, 5x5, 5x7, 5x9, 7x7, 7x9, 9x9

6. Find all the multiples where one factor is even and one factor is odd.
   1x2, 1x4, 1x6, 1x8, 2x3, 2x5, 2x7, 2x9, 3x4, 3x6, 3x8, 4x5, 4x7, 4x9, 5x6, 5x8, 6x7, 6x9, 7x8, 8x9

Number Theory for Multiplication: Multiples
Using Physical Models for Conceptual Understanding

Are you my Multiple?

Teaching tip for students: *Multiples are the answer (product) when you multiply 2 factors.*☺

(**CCSS**: 3.0A.A1; 4.0A.B.4; 6.NS.B.4)

Manipulatives Needed:
- *Multiplication books*

Preparation:
- Copy *Are you my Multiple?* worksheet for each student.
- Students will use their multiplication books created from pages 11-12.

Background concept – Multiples refer to the product or answer to a multiplication sentence.

Procedure:
- Instruct students to use their multiplication books and *Are you my Multiple?* worksheet to explore multiples.
    1) List all the multiples of …(specify number 1-9).

Example of the multiples of 3

Multiplication Table	Multiples
3	3, 6, 9, 12, 15, 18, 21, 24, 27

2) List all the multiples that appear more than twice and who they belong to (multiple of)?

Example of 12

Multiple	I am a multiple of…
12	2, 3, 4, 6

Are you my Multiple?

Use your multiplication books to:

1. List all the multiples (these are the answers to all your multiplication sentences).

Multiplication Table	Multiples
1	1, 2, 3, 4, 5, 6, 7, 8, 9,…
2	
3	
4	
5	
6	
7	
8	
9	

2. List all the multiples that appear more than twice in your book. Who do they belong to (multiple of)?

Multiple	I am a multiple of …
4	1, 2, 4

Name_____ Date_____

Are you my Multiple? **Answer sheet**

Use your multiplication books to:

1. List all the multiples (these are the answers to all your multiplication sentences).

Multiplication Table Fact	Multiples
1	1, 2, 3, 4, 5, 6, 7, 8, 9,…
2	2, 4, 6, 8, 10, 12, 14, 16, 18
3	3, 6, 9, 12, 15, 18, 21, 24, 27
4	4, 8, 12, 16, 20, 24, 28, 32, 36
5	5, 10, 15, 20, 25, 30, 35, 40, 45
6	6, 12, 18, 24, 30, 36, 42, 48, 54
7	7, 14, 21, 28, 35, 42, 49, 56, 63
8	8, 16, 24, 32, 40, 48, 56, 64, 72
9	9, 18, 27, 36, 45, 54, 63, 72, 81

2. List all the multiples that appear more than twice in your book. Who do they belong to (multiple of)?

Multiple	I am a multiple of …
4	1, 2, 4
6	1, 2, 3, 6
8	1, 2, 4, 8
9	1, 3, 9
12	2, 3, 4, 6
16	2, 4, 8
18	2, 3, 6, 9
24	3, 4, 6, 8
36	4, 6, 9

Special Note: these answers are just from their multiplication books (e.g., 36 is a multiple of 3 but won't appear in their books).

Number Theory for Multiplication: Prime & Composite
Using Physical Models for Conceptual Understanding

Prime or Composite?

Teaching tip for students: Counting the number of factors is the key to identifying a prime or composite number. ☺

(**CCSS**: 4.0A.B.4)

Manipulatives Needed:
* 20 *Base Ten blocks* (just the single units) per student or *Multiplication books* created from pages 11-12.
* *Prime or Composite – That is the question!* worksheet

Preparation:
* Provide each student with a copy of the *Prime or Composite* worksheet (page 43).
* Distribute manipulatives to individuals or groups of students. You may also use the manipulatives for class demonstration.

  Mathematical concept – A natural number that has two different factors is known as a prime number. A natural number that has more than two factors is known as a composite number. *Special note*: the number one is neither prime nor composite as it does not have 2 distinct factors. I tell students it has its own unique identity! Remember the identity property? ☺

Procedure:
* **Base Ten blocks**: Students use Base Ten blocks to build all the possible rectangle/squares for numerals 1-20. This is more effective with younger students.
* **Multiplication books** : Students use their multiplication books to identify which numbers are prime or composite according to the number of distinct arrays they have.

  Note: A prime number will only have one possible array. A composite number will have more than one array (e.g., rectangle/square shape).

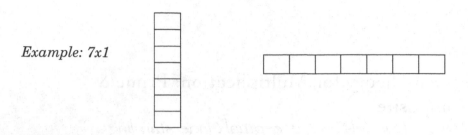

Example: 7x1

7x1 and 1x7 are the same rectangle and only have 2 distinct factors (1,7) so 7 is prime.

Example: 9

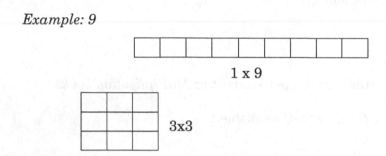

1 x 9

3x3

The numeral 9 has two distinct arrays and three factors (1, 3, 9) so it is a composite number.

- Students will record their findings on the *Prime or Composite – That is the question!* worksheet

Examples:

Prime or Composite? – that is the ?

Numeral	Number of distinct rectangles	Factors	Prime or Composite?
3	1	1, 3	Prime
4	2	1, 2, 4	Composite

Prime or Composite? – that is the ?

Number	Number of distinct rectangles	Factors	Prime or Composite?
2			
3			
4			
5			
6			
7			
8			
9			
10			
11			
12			
13			
14			
15			
16			
17			
18			
19			
20			

Name_____ Date_____

Prime or Composite? – that is the ? Answer Sheet

Numeral	Number of distinct rectangles	Factors	Prime or Composite?
2	1	1, 2	Prime
3	1	1, 3	Prime
4	2	1, 2, 4	Composite
5	1	1, 5	Prime
6	2	1, 2, 3, 6	Composite
7	1	1, 7	Prime
8	2	1, 2, 4, 8	Composite
9	2	1, 3, 9	Composite
10	2	1, 2, 5, 10	Composite
11	1	1, 11	Prime
12	3	1, 2, 3, 4, 6, 12	Composite
13	1	1, 13	Prime
14	2	1, 2, 7, 14	Composite
15	2	1, 3, 5, 15	Compsite
16	3	1, 2, 3, 8, 16	Composite
17	1	1, 17	Prime
18	3	1, 2, 3, 6, 18	Composite
19	1	1, 19	Prime
20	3	1, 2, 4, 5, 10, 20	Composite

Basic Multiplication Concepts: Games & Activities

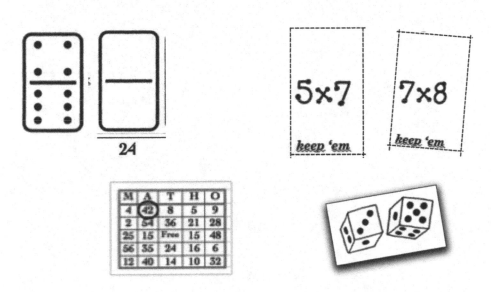

Games and activities designed to assist students in gaining procedural skills and fluency in basic multiplication concepts.

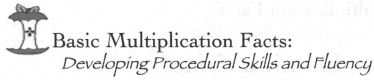 Basic Multiplication Facts:
Developing Procedural Skills and Fluency

What's my Fact?

(CCSS: 3.0A.A.1; 3.0A.A.4; 3.0A.C.7)

Manipulatives Needed:
- *Unifix™ cubes* (up to 9 colors), interlocking cubes or colored square tiles
- *Multiplying Counter* cards (1 set for each group of 2-4 students)

Preparation:
- Distribute an ample supply of cubes/tiles per group. They will be building multiplication sentences with these.
- Copy and cut apart the *Multiplying Counter* cards (pages 4-6).

Game Instructions:
- Divide the class into teams of 2-4 players and distribute manipulatives.
- Teams will place a shuffled deck of *Multiplying Counter* cards in the center of the playing area along with a pile of colored squares/cubes.
- Each player selects a *Multiplying Counter* card without showing it to the other players and constructs that multiplication sentence with their cubes/tiles. The player to the right is to guess their sentence and product. If the guess is correct, they receive that player's card. If the guess is incorrect, the player gets to keep their card.
- Game continues within a certain time limit or until all the cards have been used.

 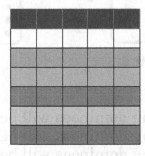

Special note: Students must construct their sentence in the order in which their card reads (e.g., 2x3 must be 2 sets of 3) and students must guess that versus saying 3x2.

 Basic Multiplication Facts:
Developing Procedural Skills and Fluency

Domino Time!

(**CCSS:** 3.OA.A.1; 3.OA.A.4; 3.OA.C.7)

Manipulatives Needed:
- Double nine set of dominoes per student
- *Domino Time* worksheets

Preparation:
- Create double nine domino sets (if you do not have commercial ones) by duplicating pages 49-50 onto cardstock and cutting them apart. Remove the dominoes that have a blank space (zero) or one dot (one). There will be 31 dominoes left. I number each set and store them in resealable bags so that they do not get mixed up.
- Duplicate *Double Time* worksheets for each student (pages 51-54).

Procedure:
- Students will use their domino sets to solve the *Domino Time* [A-C] worksheets. The dots on the dominoes represent the factors to multiply. There will be 3 dominoes left per worksheet. Some of the spaces will use 2 dominoes (e.g., 3x4 and 2x6 both represent 12).
- Students can either record their answers on the worksheets or leave the dominoes on the spaces to assess for accuracy.
- The blank *Domino Time* worksheet (page   ) can be used for students to use their domino sets to create problems for their classmates to solve.

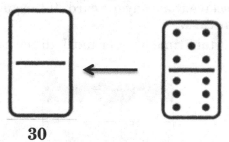

30

Answers:

[A] - The 3 remaining dominoes will be 4, 6, and 8
[B] - The 3 remaining dominoes will be 10, 32, and 81
[C] - The 3 remaining dominoes will be 21, 72, and 28

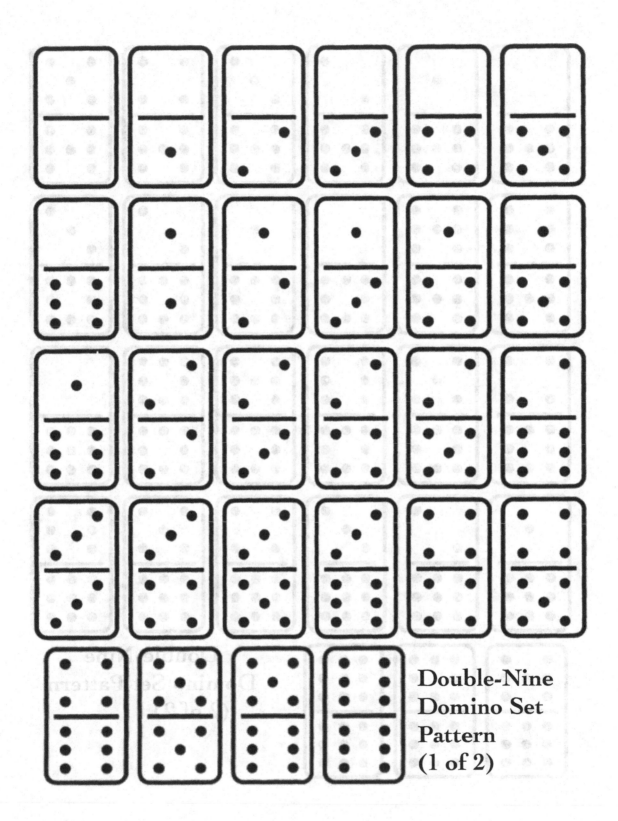

**Double-Nine
Domino Set
Pattern
(1 of 2)**

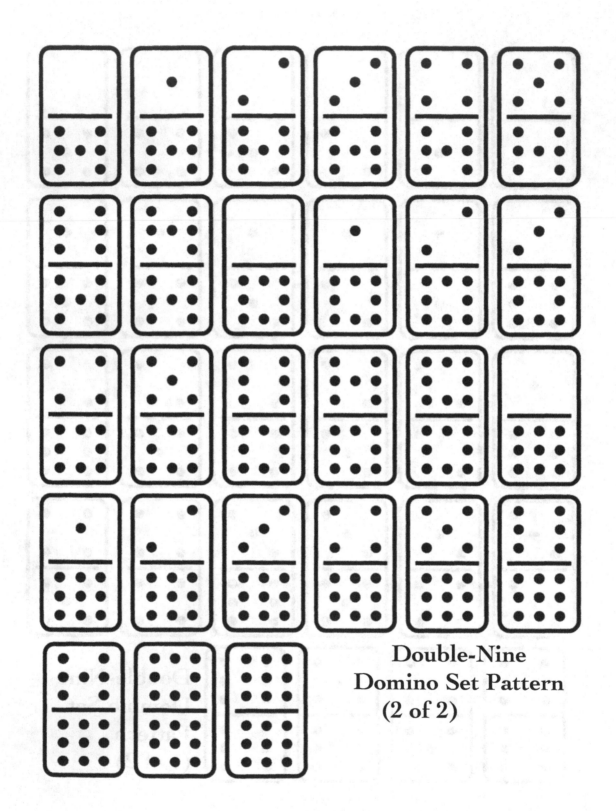

Double-Nine
Domino Set Pattern
(2 of 2)

Domino Time! [A]

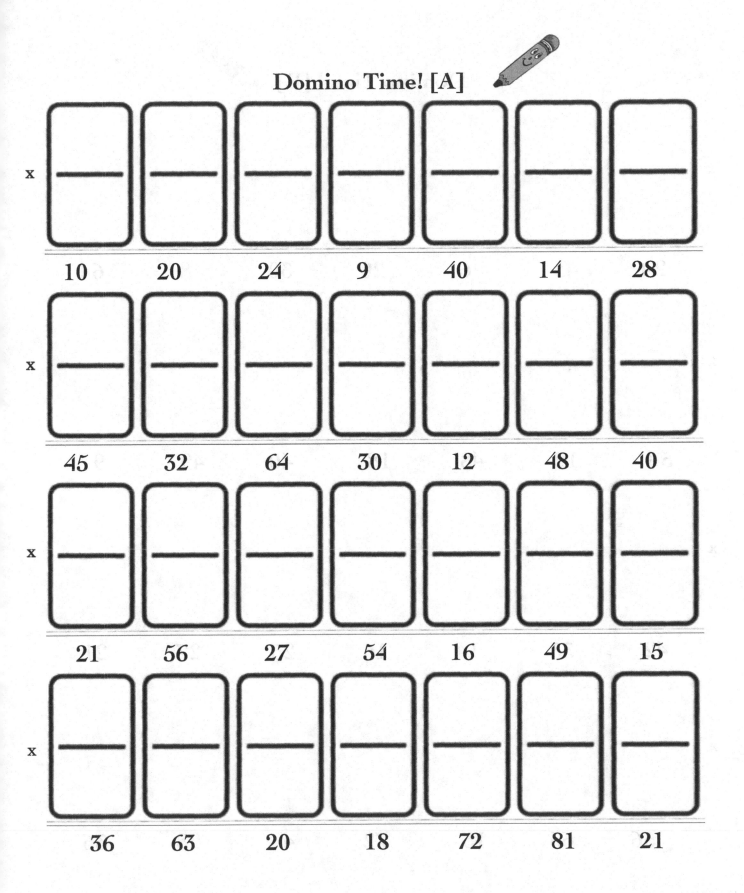

x   10   20   24   9   40   14   28

x   45   32   64   30   12   48   40

x   21   56   27   54   16   49   15

x   36   63   20   18   72   81   21

Domino Time! [B]

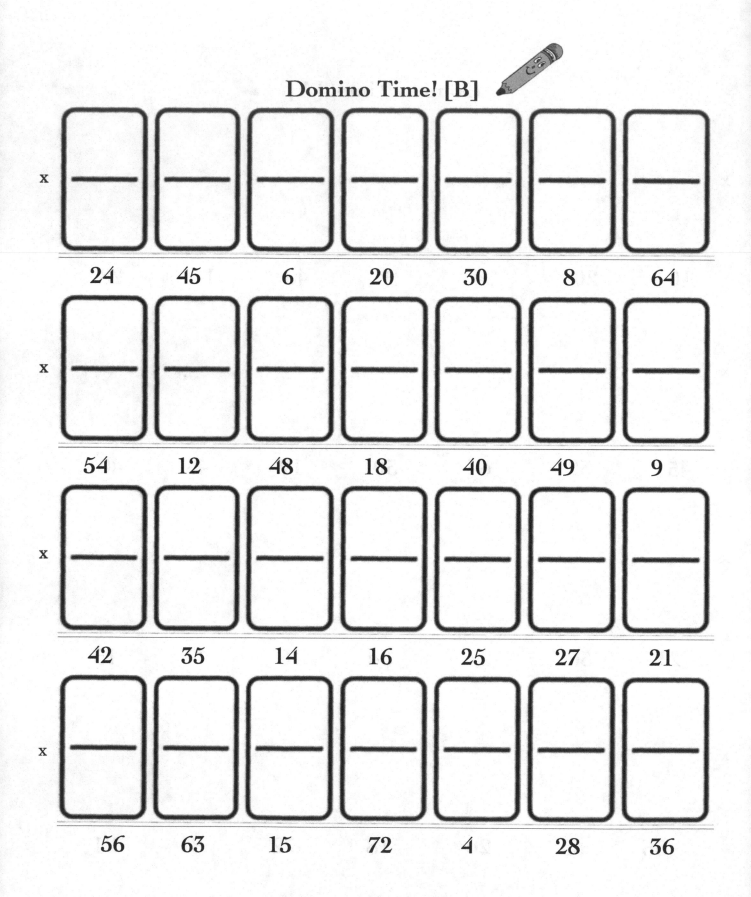

x | 24 45 6 20 30 8 64

x | 54 12 48 18 40 49 9

x | 42 35 14 16 25 27 21

x | 56 63 15 72 4 28 36

Domino Time! [C]

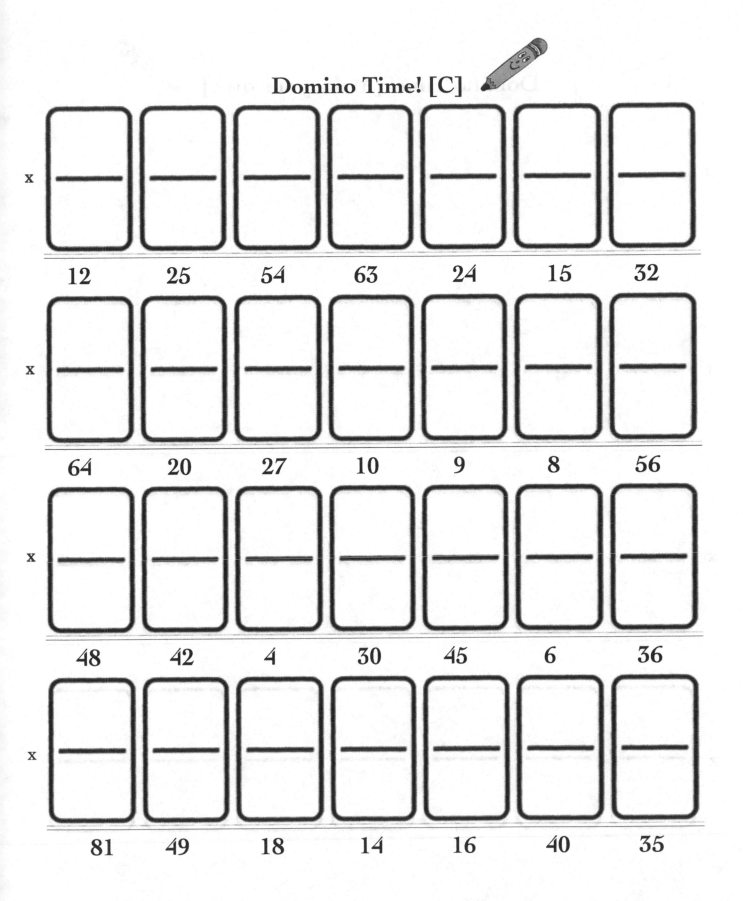

x

| 12 | 25 | 54 | 63 | 24 | 15 | 32 |

x

| 64 | 20 | 27 | 10 | 9 | 8 | 56 |

x

| 48 | 42 | 4 | 30 | 45 | 6 | 36 |

x

| 81 | 49 | 18 | 14 | 16 | 40 | 35 |

Domino Time! [Make your own]

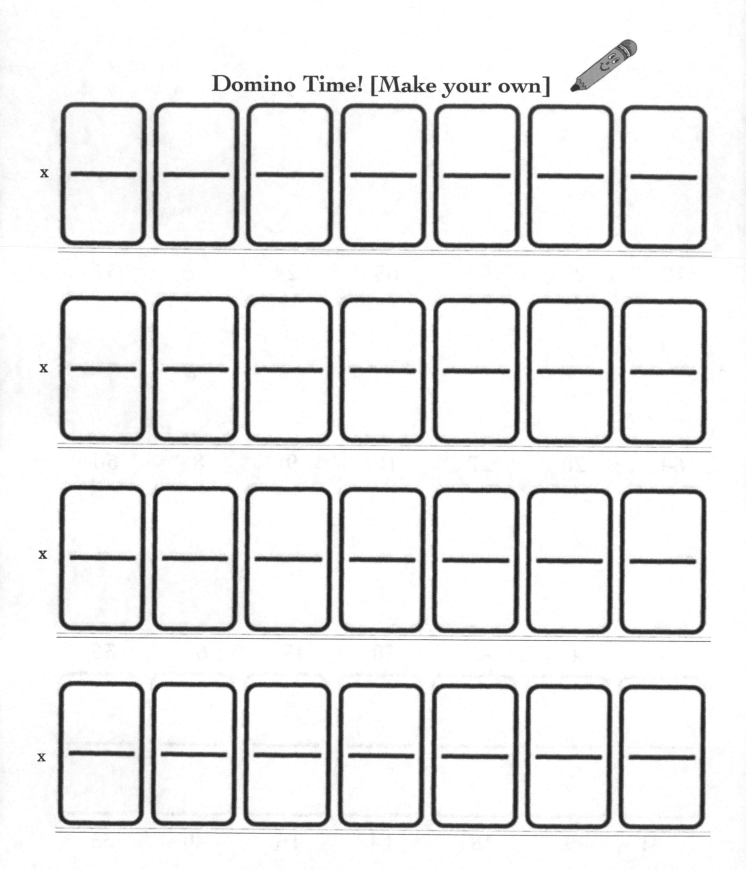

 Basic Multiplication Facts:
Developing Procedural Skills and Fluency

Guess it? Keep 'em!

(**CCSS** 3.0A.A.1; 3.0A.C.7)

Manipulatives Needed:
- *Multiplication books* from pages 11-12 (if needed)
- *Guess it? Keep'em!* cards (1 set per group of 4 students)

Preparation:
- Copy and cut apart the *Guess it? Keep 'em!* cards (pages 56-57) onto cardstock.

Game Instructions:
- Divide the class into teams of 2-4 players and distribute the manipulatives to each team.
- Teams will place a shuffled deck of *Guess it? Keep 'em!* cards in the center of the playing area.
- The first player draws three cards, placing two face up and the third one face down without looking at it. After stating the product of the first two cards, they must guess whether the third card's product is less than, in between, or greater than the either of the product of the two cards already solved.
- If the player guesses correctly, they get to keep the three cards. If the player guesses incorrectly, the cards are put in a used pile.
- Game continues until all the cards in the original pile are gone.
- The winner is the player with the most cards.

Example:

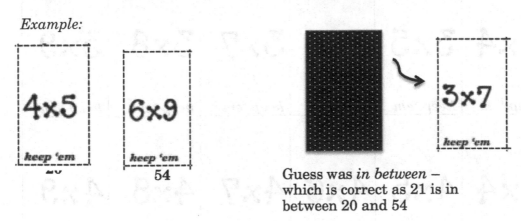

Guess was *in between* –
which is correct as 21 is in
between 20 and 54

Guess it? Keep 'em

1x1	1x2	1x3	1x4	1x5	1x6
keep 'em	keep 'em	keep 'em	keep 'em	keep 'em	keep 'em
1x7	1x8	1x9	2x2	2x3	2x4
keep 'em	keep 'em	keep 'em	keep 'em	keep 'em	keep 'em
2x5	2x6	2x7	2x8	2x9	3x3
keep 'em	keep 'em	keep 'em	keep 'em	keep 'em	keep 'em
3x4	3x5	3x6	3x7	3x8	3x9
keep 'em	keep 'em	keep 'em	keep 'em	keep 'em	keep 'em
4x4	4x5	4x6	4x7	4x8	4x9
keep 'em	keep 'em	keep 'em	keep 'em	keep 'em	keep 'em

5x5	5x6	5x7	5x8	5x9	6x6
keep 'em	keep 'em	keep 'em	keep 'em	keep 'em	keep 'em
6x7	6x8	6x9	7x7	7x8	7x9
keep 'em	keep 'em	keep 'em	keep 'em	keep 'em	keep 'em
8x8	8x9	9x9			
keep 'em	keep 'em	keep 'em			

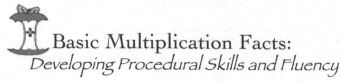 Basic Multiplication Facts:
Developing Procedural Skills and Fluency

Multi Matho

(**CCSS:** 3.0A.A.1; 3.0A.C.7)

Manipulatives Needed:
- *Counters* of any kind (e.g., buttons, beans, washers, beads, small blocks, etc...)
- Two *0-9 Spinners* per group of 2-4 players.
- *Multi Matho* game boards

Preparation:
- Copy the *Multi Math* game boards onto cardstock and cut them apart.
- Distribute the counters and spinners (page 16) to each group.

Game Instructions:
- Students begin by covering the free space on their multi math game board.
- The first player spins the two spinners which generates two factors to multiply together.
- Each player calculates the product and covers the product if it appears on their game board.
- Each player takes turns spinning and calling out the factors for the next round.
- The winner is the first play to five numbers in a straight line – MATHO!

cover

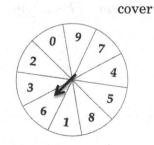

M	A	T	H	O
4	(42)	8	5	9
2	54	36	21	28
25	15	Free	15	48
56	35	24	16	6
12	40	14	10	32

Multi MATHO Game cards

M	A	T	H	O
4	42	8	5	9
2	54	36	21	28
25	15	Free	15	48
56	35	24	16	6
12	40	14	10	32

M	A	T	H	O
21	32	4	18	9
12	24	35	15	40
42	14	Free	6	28
8	48	56	10	36
16	7	49	72	64

M	A	T	H	O
6	14	10	8	12
21	18	24	4	16
9	32	Free	28	35
36	5	15	40	42
64	45	48	49	56

M	A	T	H	O
24	16	12	18	6
36	4	21	35	8
63	14	Free	10	40
9	48	28	15	3
32	42	56	27	72

M	A	T	H	O
4	42	8	5	9
2	54	36	21	28
25	15	Free	15	48
56	35	24	16	6
12	40	14	10	32

M	A	T	H	O
20	25	27	30	45
54	63	81	49	24
0	5	Free	48	7
36	18	16	8	56
21	30	9	12	64

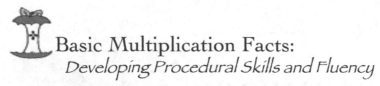

Basic Multiplication Facts:
Developing Procedural Skills and Fluency

Multiple Mascarades!

(CCSS: 3.0A.A.1; 3.0A.C.7; 6.NS.4)

Manipulatives Needed:
- *Multiple Mascarades* multiple cards
- *Multiple Mascarades* game board

Preparation:
- Copy a *Multiple Mascarades* game board for each student (page 61).
- Copy onto cardstock and cut along the dotted lines a set of *Multiple Mascarade multiple* cards for each group of students (page 62-64).

Game Instructions:
- Divide students into small groups or play game as a whole class.
- Each player places the numerals 1 through 9 (not in order) in the bottom row of squares and also along left hand column of their game board.
- Shuffle and place the *Multiple Mascarade* cards in the center of each group or for the whole class use.
- Students and/or teacher select a card and call out the multiple while each player writes that number in the corresponding square on their game board. (e.g., 49 would go in the proper square corresponding to the factors of 7x7).
- The winner is the first player whose multiples form a horizontal, diagonal, or vertical line of nine numbers.

Special note: If a multiple has more than one corresponding squares, students may fill in all of them.

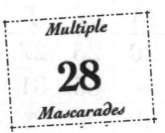

9	7	3	1	4
4	28			
7				28
X				

Multiple Mascarades game board

X								

Name_____ Date_____

Multiple Mascarades cards

Multiple	*Multiple*	*Multiple*
1	**2**	**3**
Mascarades	*Mascarades*	*Mascarades*
Multiple	*Multiple*	*Multiple*
4	**5**	**6**
Mascarades	*Mascarades*	*Mascarades*
Multiple	*Multiple*	*Multiple*
7	**8**	**9**
Mascarades	*Mascarades*	*Mascarades*
Multiple	*Multiple*	*Multiple*
10	**12**	**14**
Mascarades	*Mascarades*	*Mascarades*
Multiple	*Multiple*	*Multiple*
15	**16**	**18**
Mascarades	*Mascarades*	*Mascarades*

Multiple	*Multiple*	*Multiple*
20	**21**	**24**
Mascarades	*Mascarades*	*Mascarades*
Multiple	*Multiple*	*Multiple*
27	**28**	**30**
Mascarades	*Mascarades*	*Mascarades*
Multiple	*Multiple*	*Multiple*
32	**35**	**36**
Mascarades	*Mascarades*	*Mascarades*
Multiple	*Multiple*	*Multiple*
40	**42**	**48**
Mascarades	*Mascarades*	*Mascarades*
Multiple	*Multiple*	*Multiple*
49	**54**	**56**
Mascarades	*Mascarades*	*Mascarades*

Multiple	*Multiple*	*Multiple*
63	**64**	**72**
Mascarades	*Mascarades*	*Mascarades*
Multiple		
81		
Mascarades		

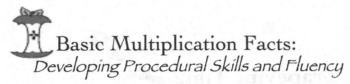

Basic Multiplication Facts:
Developing Procedural Skills and Fluency

Grapevine Time!

(**CCSS** 3.0A.A.1; 3.0A.C.7)

Manipulatives Needed:
- 2 regular *dice* per group of 2-4 students
- 1 *game marker* per student
- *Grapevine Time* game board per group of 2-4 students

Preparation:
- Copy and assemble the regular dice per instructions (page 67) if needed.
- Copy a *Grapevine Time* game board for each group (page 66).
- Distribute the game markers and dice to each group of students.

Game Instructions:
- Divide students into groups of 2-4 students each.
- The player that rolls the highest number with the 2 dice goes first.
- Players place their game marker on START.
- First player rolls both dice, multiplies the two numbers and if that product appears in the first cluster on the board, the player gets to advance their game marker to that number. If not, player can't move as they must progress sequentially through the clusters.
- Game continues with players taking turns rolling the dice, multiplying and trying to move their maker from grape cluster to grape cluster.
- If the product does not appear in the next cluster, the marker cannot advance.
- The winner is the first player to advance to the END.

Grapevine Time

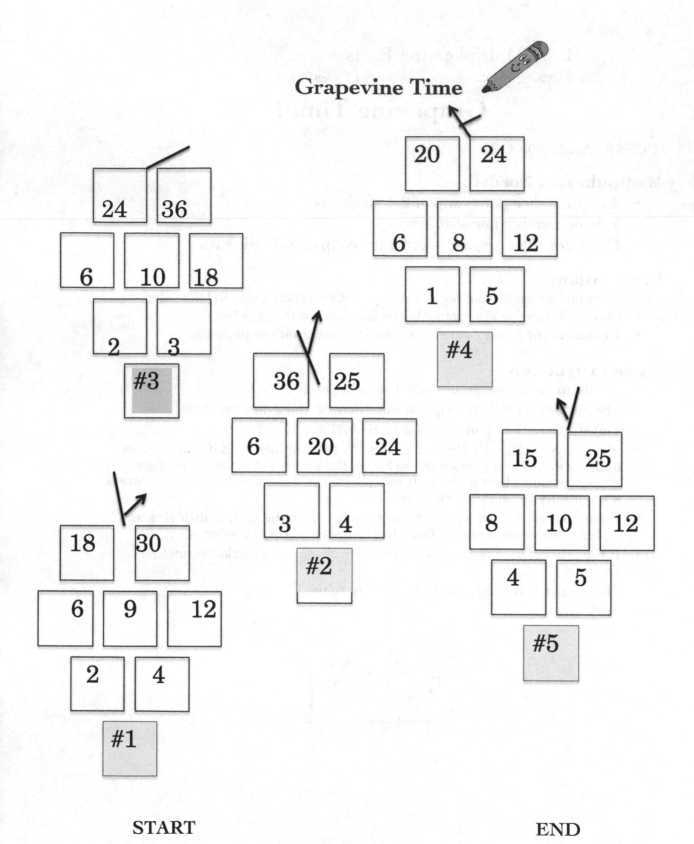

24	36

6	10	18

2	3

#3

36	25

6	20	24

3	4

#2

20	24

6	8	12

1	5

#4

15	25

8	10	12

4	5

#5

18	30

6	9	12

2	4

#1

START

END

Regular Die
[Copy onto cardstock, cut around outer edges, and glue flaps]

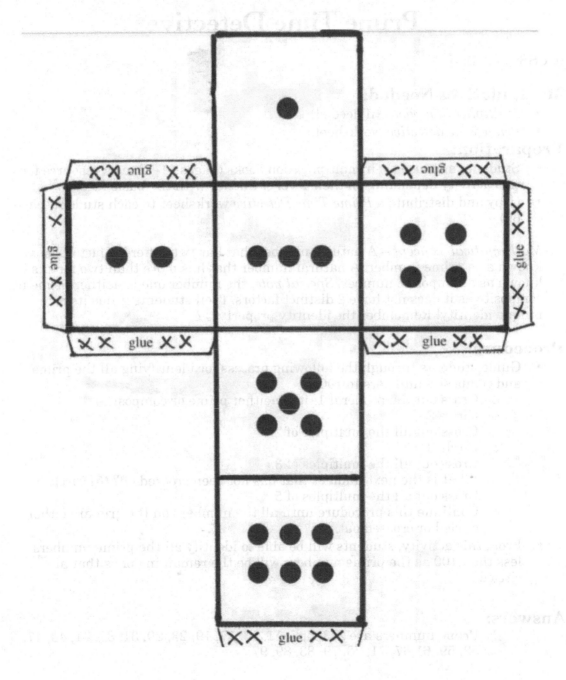

Basic Multiplication Facts:
Developing Procedural Skills and Fluency

Prime Time Detective

(CCSS: 4.0A.B.4)

Manipulatives Needed:
- *Multiplication books* (if needed)
- *Prime time detective* worksheets

Preparation:
- Students may use their multiplication books (pages 11-12) as a resource for this activity depending on their level of fluency with the basic facts.
- Copy and distribute a *Prime Time Detective* worksheet to each student (page 69).

Mathematical concept – A natural number that has two different factors is known as a prime number. A natural number that has more than two factors is known as a composite number. *Special note*: the number one is neither prime nor composite as it does not have 2 distinct factors. I tell students it has its own unique identity! Remember the identity property ☺?

Procedure:
- Guide students through the following process for identifying all the prime and composite numbers to 100:
    - Cross out the numeral 1 (it is neither prime or composite)
    - Circle 2
    - Cross out all the multiples of 2
    - Circle 3
    - Cross out all the multiples of 3.
    - What is the next number that has not been crossed off? [5] Circle 5.
    - Cross out all the multiples of 5
    - Continue this procedure until all the numbers on the grid are either circled or crossed out.
- From this activity, students will be able to identify all the prime numbers less than 100 as the prime numbers will be the remaining ones that are circled.

Answers:
Prime numbers are 2, 3, 5, 7, 11, 13, 17, 19, 23, 29, 31, 37, 41, 43, 47, 53, 59, 61,67, 71, 73, 79, 83, 89, 97.

Prime Time Detective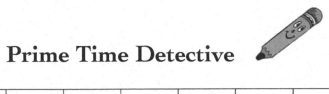

1	2	3	4	5	6	7	8	9	10
11	12	13	14	15	16	17	18	19	20
21	22	23	24	25	26	27	28	29	30
31	32	33	34	35	36	37	38	39	40
41	42	43	44	45	46	47	48	49	50
51	52	53	54	55	56	57	58	59	60
61	62	63	64	65	66	67	68	69	70
71	72	73	74	75	76	77	78	79	80
81	82	83	84	85	86	87	88	89	90
91	92	93	94	95	96	97	98	99	100

Name_____ Date_____

Multi-Digit Multiplication

165
x 43

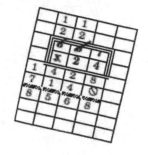

X	300	90	7
40	40 x 300	40 x 90	40x7
2	2x300	2x90	2x7

Acquiring conceptual understanding with multi-digit multiplication allows students to implement basic multiplication facts and gain fluency with strategies for complex multiplication problems.

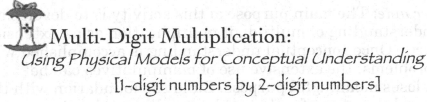

Multi-Digit Multiplication:
Using Physical Models for Conceptual Understanding
[1-digit numbers by 2-digit numbers]

Graph it: 1x2

(**CCSS:** 30A.C.7; 3.NBT.3;4.NBT.5; 5.NBT.5; 6.NS.2)

Manipulatives Needed:
- *Multi-Digit X Graph paper*
- *Graph it: 1x2* worksheet

Preparation:
- Copy 5 sheets of *Multi-Digit X* graph paper (page 75) per student.
- Copy *Graph it: 1x2* worksheet for each student.

Procedure:
- Demonstrate several 1-digit by 2-digit multiplication problems, using the graph paper to build conceptual understanding.

Example
$$\begin{array}{r} 14 \\ \times\ 7 \end{array}$$

7x4=28   +   7x10=70

28 + 70 = 98

Although this problem is stating 7 groups of 14, encourage students to process as 7 groups of 4 and 7 groups of 10 as a foundation for more complex problems.

- After adequately demonstrating a variety of problems, have students use their graph paper to illustrate the problems from their *Graph it: 1x2* worksheet.

Special teacher note: The main purpose of this activity is to develop a conceptual understanding of multi-digit multiplication as an extension of the basic facts. Once conceptual understanding is accomplished with a few larger problems, the extensive use of manipulatives can be laborious and loses meaningful purpose. Build the foundation with the manipulative and then transfer that understanding to the abstract.

Answers:

1.		1	6
	x		3
		1	8
+		3	0
		4	8

2.		2	3
	x		5
		1	5
+		1	0
	1	1	5

3.		1	4
	x		4
		1	6
+		4	0
		5	6

4.		3	4
	x		2
			8
+		6	0
		6	8

5.		5	4
	x		7
		2	8
+		3 5	0
	3	7	8

6.		1	9
	x		5
		4	5
+		5	0
		9	5

7.		4	6
	x		6
		3	6
+	2	4	0
	2	7	6

8.		3	7
	x		5
		3	5
+	1	5	0
	1	8	5

9.		1	5
	x		8
		4	0
+	8		0
	1 2		0

10.		5	8
	x		4
		3	2
+	2	0	0
	2	3	2

11.		1	8
	x		7
		5	6
+		7	0
	1	2	6

12.		6	9
	x		2
		1	8
+	1	2	0
	1	3	8

(check their graph illustrations also)

Multi-digit X Graph Paper

*Name*_____ *Date*_____

Graph it: 1x2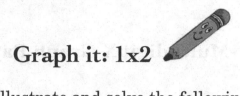

Use your graph paper to illustrate and solve the following problems.

1.
```
    1 6
  x 3
   18
  +30
   48
```

2.
```
    2 3
  x 5
```

3.
```
  1 4
  x 4
```

4.
```
    3 4
  x 2
```

5.
```
  5 4
  x 7
```

6.
```
  1 9
  x 5
```

7.
```
    4 6
  x 6
```

8.
```
  3 7
  x 5
```

9.
```
  1 5
  x 8
```

10.
```
    5 8
  x 4
```

11.
```
  1 8
  x 7
```

12.
```
  6 9
  x 2
```

Name_____ Date_____

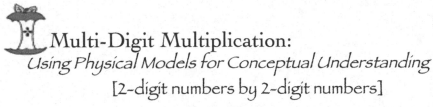

Multi-Digit Multiplication:
Using Physical Models for Conceptual Understanding
[2-digit numbers by 2-digit numbers]

Graph it: 2x2

(**CCSS**: 3.NBT.3; 4.NBT.S; 5.NBT.5; 6.NS.2)

Manipulatives Needed:
- *Multi-Digit X Graph paper*
- *Graph it: 2x2* worksheet

Preparation:
- Copy 5 sheets of graph paper (page 75) per student.
- Copy *Graph it: 2x2* worksheet for each student (page 79).

Procedure:
- Demonstrate several 2-digit by 2-digit multiplication problems, using the graph paper to build conceptual understanding.
- I have found it important to build the problems in the order of how students should be solving the problem versus the traditional method of building a large rectangle with ones, tens and hundreds. For example, instead of building 23 x17, build the thought process behind the problem (e.g., 7x3+7x20+10x3+10x20).

Example   23
             x17

10X3=30   +   10X20=200

[illustration of work from previous page]

```
      23
   x 17
      21
   +140
      161
    + 30
   + 200
      391
```

- Continue illustrating several more problems and then have students build their own using the *Multi-digit x graph paper* and the *Graph it: 2x2* worksheet.

Answers:

```
1.      4 3        2.      5 2      3.      1 3       4.       6 1
      x 2 1              x 3 2            x 4 2              x 2 2
          3                    4                6                  2
    +   4 0            + 1 0 0          +   2 0            + 1 2 0
        4 3              1 0 4              2 6              1 2 2
    +   6 0            +   6 0          + 1 2 0            +   2 0
      1 0 3              1 6 4            1 4 6              1 4 2
    + 8 0 0          1, 5 0 0          + 4 0 0          + 1, 2 0 0
      9 0 3          1, 6 6 4            5 4 6            1, 3 4 2
```

```
5.      4 2        6.      1 2      7.      3 6       8.       2 7
      x 3 4              x 7 2            x 1 6              x 2 5
          8                    4              3 6                3 5
    + 1 6 0            +   2 0          + 1 8 0            + 1 0 0
      1 6 8                2 4            2 1 6              1 3 5
    +   6 0            + 1 4 0          +   6 0            + 1 4 0
      2 2 8              1 6 4            2 7 6              2 7 5
    1, 2 0 0          + 7 0 0          + 3 0 0            + 4 0 0
    1, 4 2 8            8 6 4            5 7 6              6 7 5
```

```
9.      1 5        10     5 8      11     3 8       12      6 9
      x 5 8              x 4 4            x 2 7              x 6 2
        4 0                3 2              5 6                1 8
    +   8 0            +   2 0 0        + 2 1 0            + 1 2 0
      1 2 0              2 3 2            2 6 6              1 3 8
    + 2 5 0            + 3 2 0          + 1 6 0            + 5 4 0
      3 7 0              5 5 2            4 2 6              6 7 8
    + 5 0 0          + 2, 0 0 0        + 6 0 0          + 3, 6 0 0
      8 7 0            2, 5 5 2        1, 0 2 6            4, 2 7 8
```

Graph it: 2x2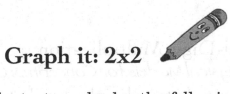

Use your graph paper to illustrate and solve the following problems.

1.   43
.   x 2 1
       3
    + 4 0
     4 3
   + 6 0
   103
  + 8 0 0
   903

2. 52
   x 3 2

3. 1 3
   x 4 2

4. 6 1
. x 2 2

5. 4 2
   x 3 4

6. 1 2
   x 7 2

7. 3 6
. x 1 6

8. 2 7
   x 2 5

9. 1 5
   x 5 8

10. 5 8
. x 4 4

11. 3 8
   x 2 7

12. 6 9
   x 6 2

Name_____ Date_____

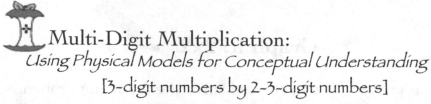

Multi-Digit Multiplication:

Using Physical Models for Conceptual Understanding

[3-digit numbers by 2-3-digit numbers]

Extending Multiplication with arrays: 3x3

(**CCSS:** 3.0A.3; 3.NBT.3; 4NBT.5; 5.NBT.5)

Manipulatives Needed:
* *Multi-Digit X Array Graph paper*
* *Array it: 3x3* worksheet

Preparation:
* Copy 4 sheets of graph paper (page 83) per student.
* Copy *Graph it 3x3* worksheet for each student (page 82).

Procedure:
Teacher note: While graph paper can be utilized to solve any problem, a lot of graph paper is required for more complex problems, thus the purpose becomes less effective. Thus, it is better to transition to arrays representing each part of the problem so that more complicated problems can be easily visualized and solved.

* Demonstrate several 3-digit by 2 to 3-digit multiplication problems, using the graph paper arrays to solve the problems.

X	300	90	7
40	40 x 300	40 x 90	40x7
2	2x300	2x90	2x7

$$
\begin{array}{r}
Example \quad 397 \\
x\ \underline{42} \\
14 \\
180 \\
+\ \underline{600} \\
794 \\
280 \\
3,600 \\
+\ \underline{12,000} \\
16,674
\end{array}
$$

- After adequately demonstrating a variety of problems, have students use their graph paper to illustrate the problems from the *Array it: 3x3* worksheet or any problems provided.

Answers:

1.
```
        4  2  3
     x  2  1
  ───────────
              3
           2  0
  +     4  0  0
  ───────────
        4  2  3
           6  0
        4  0  0
  +  8, 0  0  0
  ───────────
     8, 8  8  3
```

2.
```
        5  2  1
     x  2  5
  ───────────
              5
        1  0  0
  +  2, 5  0  0
  ───────────
     2, 6  0  5
           2  0
        4  0  0
  + 1 0, 0  0  0
  ───────────
  1 3, 0  2  5
```

3.
```
        1  3  6
     x  4  2
  ───────────
           1  2
           6  0
  +     2  0  0
  ───────────
        2  7  2
        2  4  0
     1, 2  0  0
  +  4, 0  0  0
  ───────────
     5, 7  1  2
```

4.
```
        5  1  6
     x  2  2
  ───────────
           1  2
           2  0
  +  1, 0  0  0
  ───────────
     1, 0  3  2
        1  2  0
        2  0  0
  + 1 0, 0  0  0
  ───────────
  1 1, 3  5  2
```

5.
```
        4  2  9
     x  3  6
  ───────────
           5  4
        1  2  0
  +  2, 4  0  0
  ───────────
     2, 5  7  4
        2  7  0
        6  0  0
  + 1 2, 0  0  0
  ───────────
  1 5, 4  4  4
```

6.
```
        1  2  7
     x  7  2
  ───────────
           1  4
           4  0
  +     2  0  0
  ───────────
        2  5  4
        4  9  0
     1, 4  0  0
  +  7, 0  0  0
  ───────────
     9, 1  4  4
```

7.
```
        3  6  4
     x  1  6  2
  ───────────
              8
           1  2  0
  +     6  0  0
  ───────────
        7  2  8
        2  4  0
     3, 6  0  0
  + 1 8, 0  0  0
  ───────────
  2 2, 5  6  8
           4  0  0
        6, 0  0  0
  + 3 0, 0  0  0
  ───────────
  5 8, 9  6  8
```

8.
```
        2  7  4
     x  5  8  3
  ───────────
           1  2
        2  1  0
  +     6  0  0
  ───────────
        8  2  2
        3  2  0
     5, 6  0  0
  + 1 6, 0  0  0
  ───────────
  2 2, 7  4  2
        2, 0  0  0
     3 5, 0  0  0
  + 1 0 0, 0  0  0
  ───────────
  1 5 9, 7  4  2
```

Array it: 3x3

Use your graph paper to illustrate and solve the following problems.

1.   4 2 3
  .   x 2 1

2. 5 2 1
    x 2 5

3. 1 3 6
    x 4 2

4.   5 1 6
  .   x 2 2

5. 4 2 9
    x 3 6

6. 1 2 7
    x 7 2

7.   3 6 4
  . x 1 6 2

8. 2 7 4
    x 2 5 3

Name_____ Date_____

Multi-digit X Array Paper

Name_____ Date_____

1.

2.

3.

4.

5.

9.

7.

8.

Multi-Digit Multiplication:
Using Physical Models for Conceptual Understanding

Multi-Cheers!

(**CCSS**: 4.NBT.5; 5.NBT.5; 6.NS.2)

Manipulatives Needed:
- 2 *popsicle sticks* per student
- curling ribbon, scissors, tape and a marker

Preparation:
- Instruct students to make cheerleading pom-poms by curling the ribbon and attaching it with tape to the end of each popsicle stick.
- Write *ONES* on one popsicle stick and *TENS* on the other.

- As a class create 2,3,4,5,6 and 8 syllable cheers. *For example*: (2) Cou-gars; (3) A-P-U; (4) Go –A-P-U (5) Can – you – mul-ti-ply? (6) I -love -to –mul-ti-ply.

Procedure:
Note: I developed this method to help students remember the order and direction utilized in multiplying multi-digit problems. Once these patterns are reinforced, multi-digit multiplication gains fluency. It is easiest to learn one cheer at a time and practice it before going on to another cheer.
- Students place the ONES pom-pom in the right hand and the TENS pom-pom in the left hand.
- The ONES pom-pom always represents the ones that are being multiplied. The TENS pom-pom represents the tens.
- Students raise their pom-poms in the direction and sequence of a multiplication problem while reciting the cheer. (See pattern examples on next page).
- Provide multi-digit multiplication problems for students to practice the pattern and sequence of multiplying. Repetition is the key so that the patterns are imprinted visually.
- Students now can draw the pattern on the problems and solve them easily. If they get mixed up as to what to multiply next, all they have to do is go through the pattern with their fingers.
- Introduce new patterns as they progress. Students will soon see an overall pattern and process to multiplying and will no longer need the exercise.

Variation: Sometimes older students do not want to do the cheers. Have them just use their fists to act out the patterns, then progress to only the index finger on each hand. You can also choose to have them just draw the patterns.

Sometimes cheers are too difficult for slower students. If you notice this occurring just use 1-2, 1-2-3, 1-2-3-4, etc... for the cheers.

Patterns:

[2-syllable cheer] Works with 2-digit by 1-digit problems.

With the ones pom-pom in the right hand say each syllable in the cheer with movement

Cou – gars

1st Ones in right hand straight up (Cou)
2nd Ones crosses to left (gars)

1 2
x 4

[3-syllable cheer] Works with 3-digit by 1 digit problems

sample cheer   A-P-U

1st – Ones in right hand straight up (A)
2nd – Ones to middle digit (P)
3rd – Ones to left digit (U)

259
x 6

[4-syllable cheer] Works with 2-digit by 2-digit problems or 1-digit by 3-digit

sample cheer   Go- A-P-U

1st – Ones in right hand straight up (Go)
2nd - Ones to left digit (A) –
3rd – Tens in left hand cross to upper right (P)
4th – Tens straight up (digit on left) (U)

23
x 16

[6-syllable cheer] Works with 3-digit by 2-digit problems

sample cheer   I – love – to – mul-ti-ply!

1st – Ones in right hand straight up (I)
2nd - Ones to middle digit (love) –
3rd – Ones to left digit (to)
4th – Tens in left hand to right digit (mul)
5th – Tens to middle digit (ti)
6th – Tens straight up (ply)

165
x 43

Multi-Digit Multiplication:
Developing Procedural Skills and Fluency

Managing Multi-digits

(**CCSS**: 4.NBT.5; 5.NBT.5, 6.NS.2)

Manipulatives Needed:
- *Multi-digit* graph paper (page 75)

Preparation:
- Provide ample supply of *Multi-digit* graph paper for students to show their use with any multi-digit problems they are currently solving.

Procedure:
Background information: Many of the common errors (besides computational) occur when numbers in a multi-digit multiplication problem ending up in the wrong place value space. Graph paper is an effective way to keep problems organized and preserve place value.
- Students show their work on graph paper and transfer just their answers to any type of worksheet provided. Make sure they number their problems clearly on the graph paper.

Example of what ⊘ in the spaces next place.

	1	1		
	2	2		
	3	**5**	**7**	
	x	**2**	**4**	
1	4	2	8	
7	1	4	⊘	
8	5	6	8	

their work may look like. Notice the where you are moving over to the

→ *Use spaces above for numbers carried. First row for ones, second row for tens, etc.. It is best if students cross them off once used. Once students multiply the ones, cross that off too (e.g., 4 in this example).*

 Multi-Digit Multiplication:
Developing Procedural Skills and Fluency

Estimating Products

(**CCSS**: 3.NBT.3; 4.OA.A.3; 4.NBT.5; 5.NBT.5; 6.NS.2)

Manipulatives Needed:
- *Estimating Products Spinner* game board (1 per student or group)
- *Estimating Products* record sheet (1 per student)
- 2 brads and 2 safety pins per game board

Preparation:
- Copy and assemble the *Estimating Products Spinner* game board onto cardstock (page 88).
- Provide each student with a copy of the *Estimating Products record sheet* (page 89).

Game Instructions:
- Individually or in teams, students spin both spinners to create a problem to estimate.
- Students will record spin, guess range, estimation sentence, answer, and whether their guess was correct to score a point.
- Student with the most points after the record sheet is filled wins.

   x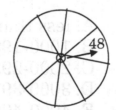

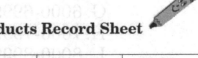

Estimating Products Record Sheet

Spin	Guess Range [A-K]	Estimation	Answer	Points
51 x 48	C	50x50	2500	1

Estimating Products Spinner Game board

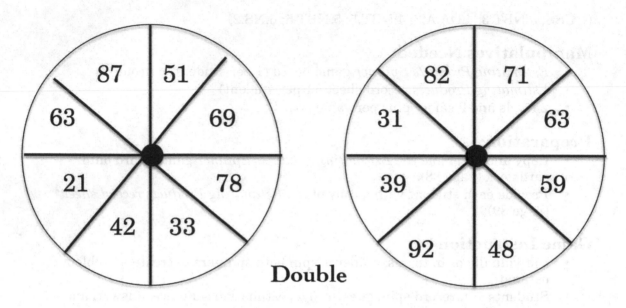

Double

Ranges...

A. less than 1000
B. 1000-1999
C. 2000-2999
D. 3000-3999
E. 4000-4999
F. 5000-5999
G. 6000-6999
H. 7000-7999
I. 8000-8999
J. 9000-9999
K. 10,000 or more

To assemble, duplicate the pattern onto cardstock. Use a brad to secure safety pin in center of spinner.

Estimating Products Record Sheet

Spin	Guess Range [A-K]	Estimation	Answer	Points

Name_____ Date_____

Multi-Digit Multiplication:
Developing Procedural Skills and Fluency

Multi-Maze!

(**CCSS**: 4.NBT.5; 5.NBT.5; 6.NS.2)

Manipulatives Needed:
- *Multi-Maze* game board (1 for each group of 4 students)
- *Multi-Maze* cards (1 for each group of 4 students)
- 15 each of two-sided counters (per student)

Preparation:
- Copy the *Multi-Maze* game board (page 92).
- Copy the *Multi-Maze* cards onto cardstock and cut them apart (page 91).

Game Instructions:
- Distribute the manipulatives to each group of players.
- Divide each group of 4 students into 2 teams.
- Place the game board in the center of the playing area with shuffled cards face down.
- Teams choose which color of the two-sided counter they will be (yellow or red).
- Team 0 draws 2 cards and multiplies them together, placing their counter color on the space that contains the answer. Cards are mixed back in with the pile.
- Team X takes their turn placing their color counter on the answer.
- Play continues until a team gets a path of answers connecting its two sides of the game board.

Multi-Maze cards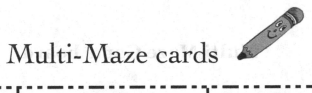

Multi-Maze	Multi-Maze	Multi-Maze
11	**31**	**51**
Multi-Maze	Multi-Maze	Multi-Maze
71	**91**	**21**
Multi-Maze	Multi-Maze	Multi-Maze
41	**61**	**81**

Multi-Maze Game board

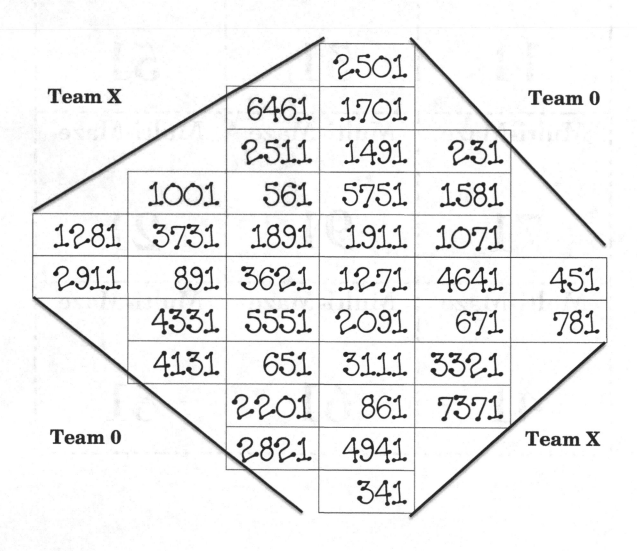

Team X

Team 0

Team 0

Team X

		2501			
	6461	1701			
	2511	1491	231		
1001	561	5751	1581		
1281	3731	1891	1911	1071	
2911	891	3621	1271	4641	451
	4331	5551	2091	671	781
	4131	651	3111	3321	
	2201	861	7371		
	2821	4941			
		341			

Basic Division Facts

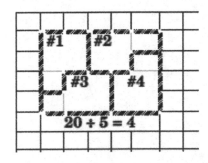

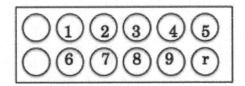

Building conceptual understanding of the basic division facts is essential in helping students gain fluency with complex division problems

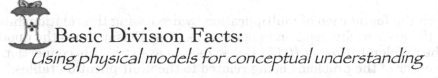

Basic Division Facts:
Using physical models for conceptual understanding

Carton It!

(**CCSS:** 3.0A.A.2; 3.0A.A.4)

Manipulatives Needed:
- Small counters of any kind (e.g., buttons, beans, washers, beads, macaroni, etc…). You will need 81 for each egg carton if you want to cover all the basic facts.
- 1 standard (holds a dozen) egg carton per student/group.
- *Carton It!* worksheets

Preparation:
- With a permanent marker write the numerals 1-9 and an "r" for remainder in the bottom of each section. The empty spaces can be used to store counters if you leave the lids on.

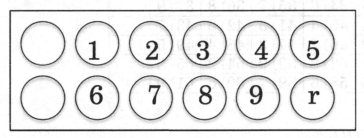

- Provide a copy of a basic division fact worksheet or *Carton It!* worksheets (pages 97-98) for each student/group.

Procedure:
- Model the process of dividing counters (dividend) into groups (divisor) with several examples until students understand the process. *For example*: Take 32 counters (this is the dividend) and divide them into 8 (divisor) sections in the carton. There will be 4 in each group (quotient).

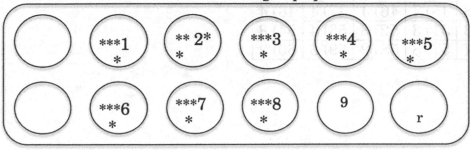

- Build upon the foundation of multiplication by discussing the relationship between the divisor, dividend and quotient, i.e., the divisor times the quotient equals the dividend. *Carton It!* [A] is designed to assist in this connection with the order of the problems being related to the multiplication tables. *Carton It!* [B] continues concept building by mixing up all the basic facts.

Special note: It is very effective to have students work in groups of 3 and trade off roles of counting out the dividend, putting counters in spaces (divisor), recording the answer (quotient).

Answers:

Carton It! [A]

1) 5	2) 6	3) 7	4) 8	5) 9	6) 2	7) 3
8) 4	9) 5	10) 6	11) 7	12) 8	13) 9	
14) 2	15) 3	16) 4	17) 5	18) 6	19) 7	
20) 8	21) 9	22) 2	23) 3	24) 4	25) 5	
26) 6	27) 7	28) 8	29) 9	30) 2	31) 3	
32) 4	33) 5	34) 6	35) 7	36) 8	37) 9	
38) 2	39) 3	40) 4	41) 5	42) 6	43) 7	
44) 8	45) 9	46) 2	47) 3	48) 4	49) 5	
50) 6	51) 7	52) 8	53) 9	54) 2	55) 3	
56) 4	57) 5	58) 6	59) 7	60) 8	61) 9	

Carton It! [B]

1) 4	2) 3	3) 5	4) 2	5) 8	6) 6
7) 3	8) 5	9) 4	10) 3	11) 8	12) 6
13) 4	14) 2	15) 7	16) 5	17) 4	18) 6
19) 5	20) 9	21) 5	22) 2	23) 7	24) 6
25) 7	26) 4	27) 5	28) 9	29) 1	30) 9
31) 7	32) 5	33) 3	34) 2	35) 7	36) 8
37) 6	38) 9	39) 8	40) 9	41) 7	42) 8
43) 5	44) 8	45) 7	46) 4	47) 4	48) 4
49) 6	50) 3	51) 3	52) 8	53) 4	54) 8
55) 9	56) 7	57) 8	58) 2	59) 6	60) 2

Carton it ! [A]

Use your egg cartons and counters to solve these problems.

1. 2)10 2. 2)12 3. 2)14 4. 2)16 5. 2)18 6. 3)6 7. 3)9

8. 3)12 9. 3)15 10. 3)18 11. 3)21 12. 3)24 13. 3)27

14. 4)8 15. 4)12 16. 4)16 17. 4)20 18. 4)24 19. 4)28

20. 4)32 21. 4)36 22. 5)10 23. 5)15 24. 5)20 25. 5)25

26. 5)30 27. 5)35 28. 5)40 29. 5)45 30. 6)12 31. 6)18

32. 6)24 33. 6)30 34. 6)36 35. 6)42 36. 6)48 37. 6)54

38. 7)14 39. 7)21 40. 7)28 41. 7)35 42. 7)42 43. 7)49

44. 7)56 45. 7)63 46. 8)16 47. 8)24 48. 8)32 49. 8)40

50. 8)48 51. 8)56 52. 8)64 53. 8)72 54. 9)18 55. 9)27

56. 9)36 57. 9)45 58. 9)54 59. 9)63 60. 9)72 61. 9)81

Carton it! [B set]

Use your egg cartons and counters to solve these problems.

1. 2)8   2. 5)15   3. 4)20   4. 8)16   5. 7)56   6. 3)18

7. 6)18   8. 2)10   9. 5)20   10. 4)12   11. 8)24   12. 7)42

13. 3)12   14. 6)12   15. 2)14   16. 5)25   17. 4)16   18. 8)48

19. 7)35   20. 3)27   21. 6)30   22. 2)4   23. 5)35   24. 4)24

25. 8)56   26. 7)28   27. 3)15   28. 6)54   29. 2)2   30. 5)45

31. 4)28   32. 8)40   33. 7)21   34. 3)6   35. 6)42   36. 2)16

37. 5)30   38. 4)36   39. 8)64   40. 9)81   41. 3)21   42. 6)48

43. 9)45   44. 4)32   45. 7)49   46. 9)36   47. 4)16   48. 8)32

49. 6)36   50. 9)27   51. 3)9   52. 5)40   53. 6)24   54. 9)72

55. 7)63   56. 9)63   57. 3)24   58. 4)8   59. 9)54   60. 9)18

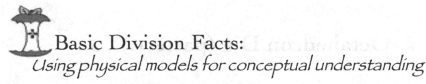

Basic Division Facts:
Using physical models for conceptual understanding

Divide those Chips!

(**CCSS**: 3.0A.A.2; 3.0A.A.4)

Manipulatives Needed:
- Small chips or counters
- 3 *Octahedron dice* per student or group of students
- *Divide those Chips!* sorting board (1 per student)
- *Divide those Chips!* record sheet (1 per student)

Preparation:
- Copy the *Octahedron die pattern* (page 100) onto cardstock and assemble according to the directions on the pattern.
- Duplicate *Divide those Chips!* sorting board (page 101) and *record sheet* (page 102) for each student or group of students.
- Distribute the manipulatives to students.

Procedure:
- Demonstrate how students will generate division problems by rolling all 3 dice. The first two dice rolled will generate the dividend and will be the amount of chips/counters needed for the *Divide those Chips!* sorting board. The third die rolled will generate the divisor (number of groups that the dividend will be sorted into).

  Example of process: Lets say a 1, 7 and 5 were rolled. Students will take 17 chips and divide them into 5 groups on their sorting board.

ᴼᴼᴼᴼᴼ	ᴼᴼᴼᴼᴼ	ᴼᴼᴼᴼᴼ
1	2	3
ᴼᴼᴼᴼᴼ 4	ᴼᴼᴼᴼᴼ 5	6
7	8	9

Remainder ᴼᴼ

Octahedron Die (pattern)

Copy onto cardstock and cut pattern out along outside edges. Fold triangles and glue the tabs to form a octahedron.

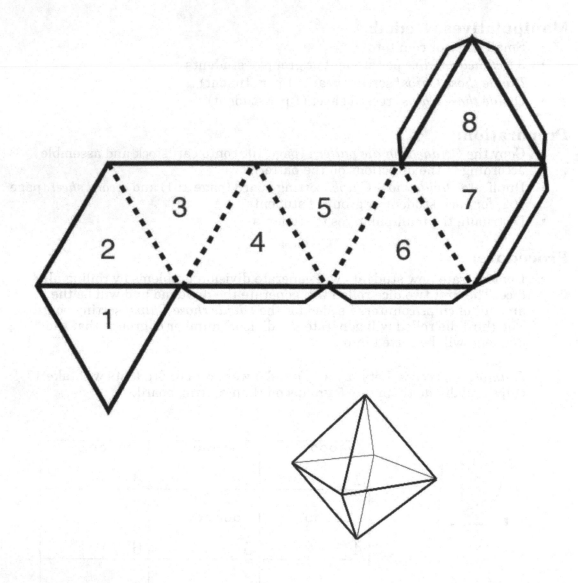

Divide those Chips! [sorting board]

1	2	3
4	5	6
7	8	9

Divide those Chips! [Record Sheet]

Record your problems and answers below.

Sentence					Remainder
1.	÷		=		
2.	÷		=		
3.	÷		=		
4.	÷		=		
5.	÷		=		
6.	÷		=		
7.	÷		=		
8.	÷		=		
9.	÷		=		
10.	÷		=		
11.	÷		=		
12.	÷		=		
13.	÷		=		
14.	÷		=		
15.	÷		=		

*Name*_____ *Date*_____

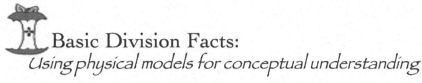 Basic Division Facts:
Using physical models for conceptual understanding

Graph Paper Facts

(**CCSS**: 3.0A.A.2; 3.0.A.A.4)

Manipulatives Needed:
- 1 cm graph paper (page 14)
- Colored pencils, crayons, or markers (2 different colors per student)
- Basic division fact worksheets (if desired)

Preparation:
- Copy and distribute graph paper (the amount of sheets of graph paper depends on the number of problems you want students to model) and the colored pencils/crayons.
- This activity can be used with any basic division fact worksheets.

Procedure:
- Demonstrate how students will use their graph paper to solve basic division facts.
  Example: 20 ÷ 5 = 4
    - o Draw a boundary around a set of 20 squares with 1 color.
    - o With the second color draw squares around groups of 5 squares.
    - o Number the groups of squares. (#1,2,3,4)
    - o Write the division sentence below the square.

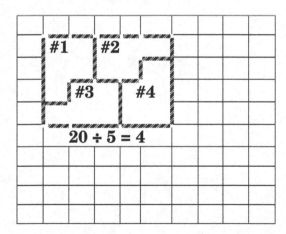

- Students can use graph paper right along with a division fact worksheet of choice, using the graph paper to show work and the worksheet to record answers.

Number Theory for Division

Divisibility is the foundational skill that enables students to determine if a number is divisible by another number empowering work with more complex processes (e.g., working with prime and composite numbers, reducing fractions and algebraic equations).

Divisibility Rules For 2,4,5,8,10
Using physical models for conceptual understanding

How does it End?

(**CCSS**: 3.0A.A.2; 3.0A.5; 3.0A.6; 4.NBT.A.1)

Manipulatives Needed:
- *Counters* of any kind (e.g., buttons, beans, washers, beads, small blocks, etc...)
- Small bowls or containers for counters
- *How does it End?* worksheets (pages 111-115)
- *Fact Finder* (pages 19-20) [for divisibility by 4]
- *Calculator*

Preparation:
- Select the divisibility rule you want your students to explore and make copies of the appropriate worksheet.
  - o Divisibility by 2 : Distribute the counters and containers to individual or groups of students. Provide copies of *How does it End? [Divisibility by 2]* worksheet (page 111).
  - o Divisibility by 4: Have students use their *Fact Finder* manipulative from pages 19-20 or provide a multiple chart. Provide copies of the *How does it End? [Divisibility by 4]* worksheet (page 112).
  - o Divisibility by 5: Students can use a calculator unless they need a greater challenge. Provide copies of the *How does it End [Divisibility by 5]* worksheet (page 113).
  - o Divisibility by 8: I prefer students not to use a calculator with this one so they stretch their mental math skills. Provide copies of the *How does it End? [Divisibility by 8]* worksheet (page 114).
  - o Divisibility by 10: Students can choose whether they want to use a calculator or not. Provide copies of the *How does it End? [Divisibility by 10]* worksheet (page 115).

Procedure:
- Demonstrate how students will use the manipulative and their worksheet to explore divisibility and solve the mystery of the rule.
- Utilize the worksheets and manipulative to facilitate conceptual understanding of the divisibility rules.

(continued on next page)

Mathematical concept – When a number is divisible by another number there is no remainder: Divisibility Rules for:

        2 – the last digit is even (0,2,4,6,8)
        4 – the last 2 digits are a multiple of 4
        5 – the last digit is 0 or 5
        8 – the last 3 digits are divisible by 8
        10 – the last digit is 0

ANSWERS to Divisibility by 2 & 4

How does it End?
[Divisibility by 2]

Use your counters and 2 containers to solve the mystery.

Number of counters to take	Divisible by 2? No	Divisible by 2? Yes	What is the last digit?
11	✔		1
12		✔	2
13	✔		3
14		✔	4
15	✔		5
16		✔	6
17	✔		7
18		✔	8
19	✔		9
20		✔	0

Circle the last digits on your chart that ended up being a Yes.
Clue to solving the mystery! If a number is divisible by 2, the last digit has to be **0, 2, 4, 6, and 8** –all even numbers!

How does it End?
[Divisibility by 4]

Use your *Fact Finder* or *Multiples of 4 chart* & a calculator to solve the mystery.

Underline last 2 digits of the mystery number	Multiple of 4? No	Multiple of 4? Yes	If yes, test with a calculator
4<u>24</u>		Yes	✔
3,2<u>95</u>	No		
4,6<u>16</u>		Yes	✔
25,5<u>36</u>		Yes	✔

4,298	No		
65,220		Yes	✔
104,204		Yes	✔
23,512		Yes	✔
19	No		
20			

Clue to solving the mystery! If a number is divisible by 4, the last 2 digits are __divisible__ of 4.

ANSWERS to Divisibility by 5, 8

How does it End?
[Divisibility by 5]

Use your *calculator* to solve the mystery.

Mystery Number	Divisible by 5? No	Divisible by 5? Yes	If yes, what is the last digit?
6,007	✔		
65,412	✔		
23,890		✔	0
41,786	✔		
215,909	✔		
237,901	✔		
470,065		✔	5
564,383	✔		
19,434	✔		
154,688	✔		

Clue to solving the mystery! If a number is divisible by 5, the last digit is a __0__ or a __5.__

How does it End?
[Divisibility by 8]

Solve the mystery. Challenge yourself without a calculator!.

Mystery Number	Write last 3 numbers	Divide by 8	Is there a remainder?	Is it divisible by 8?
8,512	512	64	No	Yes
5,037	037	4r5	Yes	No
6,854	854	106r6	Yes	No
2,752	752	94	No	Yes
5,271	271	33r7	Yes	No
7,312	312	39	No	Yes

6,328	328	41	No	Yes
7,072	072	9	No	Yes
1,578	578	72r2	Yes	No
5,763	763	95r3	Yes	No

Clue to solving the mystery! If a number is divisible by 8, the last 3 digits are
___**divisible by 8**_____.

ANSWERS to Divisibility by 10

How does it End?
[Divisibility by 10]

Use your *calculator* to solve the mystery. Challenge yourself without one!

Mystery Number	Divisible by 10? No	Divisible by 5? Yes	If yes, what is the last digit?
54,796	✔		
45,697	✔		
23,893	✔		
31,780		✔	0
405,909	✔		
438,901	✔		
470,062	✔		
986,584	✔		
564,380		✔	0
619,438	✔		
234,680		✔	
8,455	✔		0
436,850		✔	

Clue to solving the mystery! If a number is divisible by 10, the last digit is a
_0___ or the last 2 digits are divisible by 10. Go back and look at the last 2 digits.
Does it work? _**Yes**____

How does it End?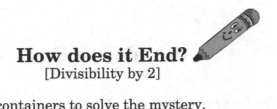
[Divisibility by 2]

Use your counters and 2 containers to solve the mystery.

Number of counters to take	Divisible by 2? No	Divisible by 2? Yes	What is the last digit?
11	X		1
12			
13			
14			
15			
16			
17			
18			
19			
20			

Circle the last digits on your chart that ended up being a Yes.
Clue to solving the mystery! If a number is divisible by 2, the last digit has to be
____. ___, ____, ____, ____, all even numbers!

How does it End?

[Divisibility by 4]

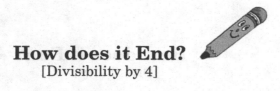

Use your *Fact Finder* or *multiples of 4 chart* & a calculator to solve the mystery.

Underline last 2 digits of the mystery number	Multiple of 4? No	Multiple of 4? Yes	If yes, test with a calculator
4<u>24</u>		Yes	✔
3,295			
4,616			
25,536			
4,298			
65,220			
104,204			
23,512			
19			
20			

Clue to solving the mystery! If a number is divisible by 4, the last 2 digits are a_____ of 4.

How does it End?
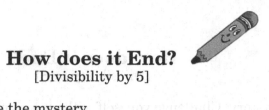
[Divisibility by 5]

Use your *calculator* to solve the mystery.

Mystery Number	Divisible by 5? No	Divisible by 5? Yes	If yes, what is the last digit?
6,007	✔		
65,412			
23,890			
41,786			
215,909			
237,901			
470,065			
564,383			
19,434			
154,688			

Clue to solving the mystery! If a number is divisible by 5, the last digit is a ____ or a ____.

How does it End?

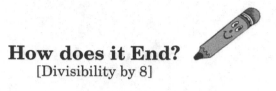

[Divisibility by 8]

Solve the mystery. Challenge yourself without a calculator!.

Mystery Number	Write last 3 numbers	Divide by 8	Is there a remainder?	Is it divisible by 8?
8,512	512	64	No	Yes
5,037				
6,854				
2,752				Y
5,271				
7,312				Y
6,328				Y
7,072				Y
1,578				
5,763				

Clue to solving the mystery! If a number is divisible by 8, the last 3 digits are _____.

How does it End?

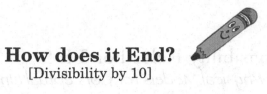

[Divisibility by 10]

Use your *calculator* to solve the mystery. Challenge yourself without one!

Mystery Number	Divisible by 10? No	Divisible by 5? Yes	If yes, what is the last digit?
54,796	✔		
45,697			
23,893			
31,780			
405,909			
438,901			
470,062			
986,584			
564,380			
619,438			
234,680			
8,455			
436,850			

Clue to solving the mystery! If a number is divisible by 10, the last digit is a ____ or the last 2 digits are divisible by 10. Go back and look at the last 2 digits. Does it work? _____

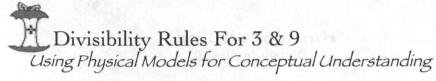

Divisibility Rules For 3 & 9

Using Physical Models for Conceptual Understanding

Sum Fun!

(**CCSS**: 3.0A.A.2; 3.0A.5; 3.0A.6; 4.NBT.A.1)

Manipulative Needed:
- *Sum Fun!* worksheets
- Multiplication books (pages 11-12)

Preparation:
- Select the divisibility rule you want your students to explore and make copies of the appropriate worksheet (pages 118-119).
- .
  - o Divisibility by 3: Students can use their multiplication books as a reference for the multiples of 3.
  - o Divisibility by 9: Students can use their multiplication books as a reference for the multiples of 9.

Procedure:
- Demonstrate how students will use the manipulative and their worksheet to explore divisibility and solve the mystery of the rule.
- Utilize the worksheets and manipulative to facilitate conceptual understanding of the divisibility rules.

Mathematical concept – When a number is divisible by another number there is no remainder:

Divisibility Rules for:

3 – the sum of the digits is a multiple of 3

9 – the sum of the digits is divisible by 9

Number	Sum of digits
327	**3+2+7=12**

Is it a multiple? [look on the 3's page]
3x4=12 (yes)

Answers:

Sum Fun!
[Divisibility by 3]

Use your addition skills and your multiplication books to check for multiples.

Mystery number	Sum of the digits	Is it a multiple of 3?	Is it divisible by 3?
4,696	24	Yes	Yes
5,692	22	No	No
102,132	9	Yes	Yes
123,456	21	Yes	Yes
798,471	36	Yes	Yes
102,948	23	No	No
1,023	6	Yes	Yes
42,561	18	Yes	Yes
3,655	19	No	No
17,845	17,845	No	No

Clue to solving the mystery! If a number is divisible by 3, the **SUM** of the digits is
a **multiple** of 3.

Sum Fun!
[Divisibility by 9]

Use your addition skills and your multiplication books to check for multiples.

Mystery number	Sum of the digits	Is it a multiple of 9?	Is it divisible by 9?
6,984	27	Yes	Yes
2,375	17	No	No
975,645	36	Yes	Yes
10,233	9	Yes	Yes
647,382	30	No	No
479,988	45`	Yes	Yes
394,958	38	No	No
989,991	54	Yes	Yes
236,438	26	No	No
548,439	33	No	No

Clue to solving the mystery! If a number is divisible by 9, the **SUM** of the
digits is a **MULTIPLE** of 9 or divisible by 9.

Sum Fun!

[Divisibility by 3]

Use your addition skills and your multiplication books to solve the mystery.

Mystery number	Sum of the digits	Is it a multiple of 3?	Is it divisible by 3?
4,696	24	Yes	Yes
5,692			
102,132			
123,456			
798,471			
102,948			
1,023			
42,561			
3,655			
17,845			

Clue to solving the mystery! If a number is divisible by 3, the _____ of the digits is

a _____ of 3 or divisible by 3.

Sum Fun!

[Divisibility by 9]

Use your addition skills and your multiplication books to solve the mystery.

Mystery number	Sum of the digits	Is it a multiple of 9?	Is it divisible by 9?
6,984	27	Yes	Yes
2,375			
975,645			
10,233			
647,382			
479,988			
394,958			
989,991			
236,438			
548,439			

Clue to solving the mystery! If a number is divisible by 9, the _____ of the digits is

a _____ of 9 or divisible by 9.

Divisibility Rules For 6 & 7

Using physical models for conceptual understanding

It's Complicated!

(**CCSS**: 3.0A.A.2; 3.0A.5; 3.0A.6; 4.NBT.A.1)

Manipulatives Needed:
- *It's Complicated!* worksheets
- Multiplication books (pages 11-12)
- Calculator (optional)

Preparation:
- Select the divisibility rule you want your students to explore and make copies of the appropriate worksheet (pages 122-124).
  - Divisibility by 6: Students can use their multiplication books as a reference for the multiples of 2 and 3
  - Divisibility by 7: Students can use their multiplication books as a reference for the multiples of 7. Calculators can be used for doubling and subtraction.

Procedure:
- Demonstrate how students will use the manipulative and their worksheet to explore divisibility and solve the mystery of the rule.
- Utilize the worksheets and manipulative to facilitate conceptual understanding of the divisibility rules.

Mathematical concept – When a number is divisible by another number there is no remainder:

Divisibility Rules for:

6 – the number has to pass the divisibility test for both 2 and 3.

2 – last digit ends in 0,2,4,6 or 8

3 – sum of the digits is a multiple of 3

7 – Multiply the last digit by 2 (or double it). Subtract that number from the remaining numbers and check if it is divisible by 7. You may need to repeat the steps if you don't recognize the number as divisible by 7 yet. If the last digit is 0, double the next number to the left (see it's complicated, but fun!).

Example of divisibility by 7: 245

-10 (this is 5 doubled)

14 – this is divisible by 7!

Answers:

It's Complicated!
[Divisibility by 6]

Use your 2 & 3 divisibility rules to solve the mystery.

Mystery number	Is it divisible by 2? (ends in 0,2,4,6,8)	Is it divisible by 3 (sum of digits multiple of 3)	Is it divisible by 2 & 3?
12	Yes	3/yes	✔
456	Yes	15/yes	✔
289	No	19/no	
692	Yes	17/n0	
862	Yes	16/no	
2,568	Yes	21/yes	✔
4,698	Yes	27/yes	✔
1,264	Yes	13/no	
3,966	Yes	24/yes	✔
5,566	Yes	22/no	

Clue to solving the mystery! If a number is divisible by 6, it is divisible by __2__ and __3__.

It's Complicated!
[Divisibility by 7]

Use the math steps given to solve the mystery.

Mystery number	Double last digit	Subtract from remaining number	Is it remaining number divisible by 7?
368	8x2=16	36-16=**20**	No
245	5x2 = 10	24-10=**14**	Yes
154	4x2=8	15-8=7	Yes
203	3x2=6	20-6=14	Yes
133	3x2=6	13-6=7	Yes
556	6x2=12	55-12=43	No
4,698	8x2=16	469-16=453/45-6=39	No
3,192	2x2=4	319-4=315/31-10=21	Yes
444	4x2=8	44-8=36	No
38,241	1x2=2	3824-2=23822-4=378-16=21	Yes

Clue to solving the mystery! If a number is divisible by 7, __double__ the last digit, __Subtract__ from the remaining number (repeat if necessary) to see if remaining number is **divisible** by 7

It's Complicated!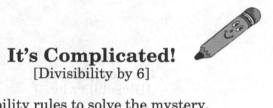

[Divisibility by 6]

Use your 2 & 3 divisibility rules to solve the mystery.

Mystery number	Is it divisible by 2? (ends in 0,2,4,6,8)	Is it divisible by 3? (sum of digits multiple of 3)	Is it divisible by 2 & 3?
12	Yes	3/yes	✔
456			
289			
692			
862			
2,568			
4,698			
1,264			
3,966			
5,566			

Clue to solving the mystery! If a number is divisible by 6, it is divisible by ____ and ____.

It's Complicated!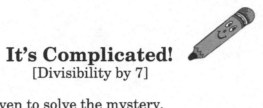
[Divisibility by 7]

Use the math steps given to solve the mystery.

Mystery number	Double last digit	Subtract from remaining number	Is it remaining number divisible by 7?
368	8x2=**16**	36-16=**20**	No
245	5x2 = **10**	24-10=**14**	Yes
154			
203			
133			
556			
4,698			
3,192			
444			
38,241			

Clue to solving the mystery! If a number is divisible by 7, ___ the last digit, _____ from the remaining number (repeat if necessary) to see if remaining number is ___ by 7.

Basic Division Concepts: Games & Activities

 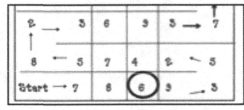 

Games and activities designed to assist students in gaining procedural skills and gaining fluency in basic division concepts.

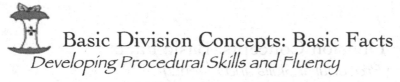

Basic Division Concepts: Basic Facts
Developing Procedural Skills and Fluency

Sharing is Caring!

(**CCSS**: 3.OA.A.2; 3.OA.A.4; 3.OA.C.7)

Manipulatives Needed:
- 25 counters per student
- 3 regular dice per group (page 67)
- Cup/container for each student and one for the group
- Paper and pencil to keep score of points

Preparation:
- Divide students into groups of 3, 4 or 5 players.
- Distribute game manipulatives to each student.

Game Instructions:
- Each player starts with 25 counters in their cup/container. The player who rolls the highest sum of 3 dice goes first. The first player may choose to roll 2 or 3 dice, find the sum, and take that many counters out of their cup/container. This player then shares (divides) these counters with the other players so that each receives the same amount.
- If there are counters left, they are placed in a cup/container in the center of the playing area.
- Play continues with the next player. If a player does not have enough chips to share, they pass to the next player.
- The first player to get rid of all their counters and go out exactly collects all the counters in the center container and receives 1 point for each.
- Game starts over again.
- The first player to reach 50 points wins.

Example: Player chooses 2 dice and rolls a 5 and a 6 and takes 11 counters.

If there are 4 players in this group, each gets 2 counters and player puts the remaining 3 in the middle containers.

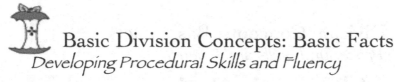

Basic Division Concepts: Basic Facts
Developing Procedural Skills and Fluency

Divide & Conquer!

(**CCSS**: 3.OA.A.2; 3.OA.A.4; 3.OA.C.7))

Manipulatives Needed:
- *Divide & Conquer* playing cards

Preparation:
- Copy and cut out the *Divide and Conquer* playing cards (pages 129-134). It helps to number them and store them in re-sealable bags so that the sets don't get mixed up.
- Distribute 1 set of *Divide and Conquer* playing cards to each set of two students.

Game Instructions:
- This game is played like the traditional card game called War.
- Divide students into teams of two players.
- Shuffle the cards and deal 32 cards face down to each player.
- Each player turns over their top card and states the answer (quotient). Whoever has the highest quotient wins both cards.
- If there is a tie, war is declared and another set of cards are turned over. Whoever has the highest answer (quotient) takes all 4 cards.
- When a player has played all his cards, they are shuffled and used again.
- Play continues until one player has lost all his cards to the other player who becomes the winner.

Special note: If a player gives an incorrect answer, other player automatically earns the card.

Player 1

Player 2

30 ÷ 5=

Divide & Conquer

27 ÷ 3=

Divide & Conquer

5

7

Player 2 takes both cards

2 ÷ 2 =

Divide & Conquer

4 ÷ 2 =

Divide & Conquer

6 ÷ 2 =

Divide & Conquer

8 ÷ 2 =

Divide & Conquer

10 ÷ 2 =

Divide & Conquer

12 ÷ 2 =

Divide & Conquer

14 ÷ 2 =

Divide & Conquer

16 ÷ 2 =

Divide & Conquer

18 ÷ 2 =

Divide & Conquer

3 ÷ 3 =

Divide & Conquer

6 ÷ 3 =

Divide & Conquer

9 ÷ 3 =

Divide & Conquer

12 ÷ 3 =	**15 ÷ 3 =**	**18 ÷ 3 =**
Divide & Conquer	Divide & Conquer	Divide & Conquer
21 ÷ 3 =	**24 ÷ 3 =**	**27 ÷ 3 =**
Divide & Conquer	Divide & Conquer	Divide & Conquer
4 ÷ 4 =	**8 ÷ 4 =**	**12 ÷ 4 =**
Divide & Conquer	Divide & Conquer	Divide & Conquer
16 ÷ 4 =	**20 ÷ 4 =**	**24 ÷ 4 =**
Divide & Conquer	Divide & Conquer	Divide & Conquer

28 ÷ 4 =	**32 ÷ 4 =**	**36 ÷ 4 =**
Divide & Conquer	Divide & Conquer	Divide & Conquer
5 ÷ 5 =	**10 ÷ 5 =**	**15 ÷ 5 =**
Divide & Conquer	Divide & Conquer	Divide & Conquer
20 ÷ 5 =	**25 ÷ 5 =**	**30 ÷ 5 =**
Divide & Conquer	Divide & Conquer	Divide & Conquer
35 ÷ 5 =	**40 ÷ 5 =**	**45 ÷ 5 =**
Divide & Conquer	Divide & Conquer	Divide & Conquer

$6 \div 6 =$	$12 \div 6 =$	$18 \div 6 =$
Divide & Conquer	Divide & Conquer	Divide & Conquer
$24 \div 6 =$	$30 \div 6 =$	$36 \div 6 =$
Divide & Conquer	Divide & Conquer	Divide & Conquer
$42 \div 6 =$	$48 \div 6 =$	$54 \div 6 =$
Divide & Conquer	Divide & Conquer	Divide & Conquer
$7 \div 7 =$	$14 \div 7 =$	$21 \div 7 =$
Divide & Conquer	Divide & Conquer	Divide & Conquer

28 ÷ 7 =	**35 ÷ 7 =**	**42 ÷ 7 =**
Divide & Conquer	Divide & Conquer	Divide & Conquer
49 ÷ 7 =	**56 ÷ 7 =**	**63 ÷ 7 =**
Divide & Conquer	Divide & Conquer	Divide & Conquer
8 ÷ 8 =	**16 ÷ 8 =**	**24 ÷ 8 =**
Divide & Conquer	Divide & Conquer	Divide & Conquer
32 ÷ 8 =	**40 ÷ 8 =**	**48 ÷ 8 =**
Divide & Conquer	Divide & Conquer	Divide & Conquer

$56 \div 8 =$	$64 \div 8 =$	$72 \div 8 =$
Divide & Conquer	Divide & Conquer	Divide & Conquer
$9 \div 9 =$	$18 \div 9 =$	$27 \div 9 =$
Divide & Conquer	Divide & Conquer	Divide & Conquer
$36 \div 9 =$	$45 \div 9 =$	$54 \div 9 =$
Divide & Conquer	Divide & Conquer	Divide & Conquer
$63 \div 9 =$	$72 \div 9 =$	$81 \div 9 =$
Divide & Conquer	Divide & Conquer	Divide & Conquer

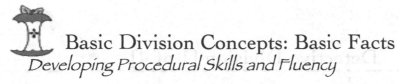
Detective Division

(**CCSS:** 3.0A.A.2; 3.0A.A.4; 3.0A.B.6)

Manipulatives Needed:
- 81 *Base-ten unit* blocks per group of 4 students
- *Detective Division* playing cards (1 set per group)

Preparation:
- Copy and cut out the *Detective Division* playing cards (pages 136-139). It helps to number them and store them in re-sealable bags so that the sets don't get mixed up.
- Distribute 1 set of cards and 81 base ten unit blocks to each group of students.

Game Instructions:
- Divide students into groups and instruct them to shuffle the cards and place them face down in the center of their playing area
- Each player takes a turn picking up a card (don't show it to the other players) and constructs the division problem with their blocks. If a player picks a *"Build your own"* card they may construct any basic division fact sentence.
- The other players watch to figure out what division problem they are illustrating but only the player to the right is to guess their sentence and quotient.
- It is important that students guess the sentence in the right order.

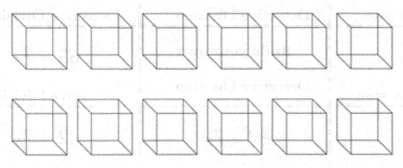

i.e. $12 \div 6 = 2$

- If their detective work is correct, they receive the player's card. If incorrect, the player gets to keep the card.
- Play continues around the group until all cards have been built. If you want to speed the game up, students can all build their division problems at the same time.

81 ÷ 9 =	72 ÷ 9 =	63 ÷ 9 =
Detective Division	Detective Division	Detective Division
54 ÷ 9 =	45 ÷ 9 =	36 ÷ 9 =
Detective Division	Detective Division	Detective Division
27 ÷ 9 =	18 ÷ 9 =	9 ÷ 9 =
Detective Division	Detective Division	Detective Division
Build your own Detective Division	72 ÷ 8 =	64 ÷ 8 =
	Detective Division	Detective Division
56 ÷ 8 =	48 ÷ 8 =	40 ÷ 8 =
Detective Division	Detective Division	Detective Division
32 ÷ 8 =	24 ÷ 8 =	16 ÷ 8 =
Detective Division	Detective Division	Detective Division
8 ÷ 8 =	**Build your own** Detective Division	63 ÷ 7 =
Detective Division		Detective Division
56 ÷ 7 =	49 ÷ 7 =	42 ÷ 7 =
Detective Division	Detective Division	Detective Division
35 ÷ 7 =	28 ÷ 7 =	21 ÷ 7 =
Detective Division	Detective Division	Detective Division

14 ÷ 7 = Detective Division	7 ÷ 7 = Detective Division	**Build your own** Detective Division
54 ÷ 6 = Detective Division	48 ÷ 6 = Detective Division	42 ÷ 6 = Detective Division
36 ÷ 6 = Detective Division	30 ÷ 6 = Detective Division	24 ÷ 6 = Detective Division
18 ÷ 6 = Detective Division	12 ÷ 6 = Detective Division	6 ÷ 6 = Detective Division
Detective Division	45 ÷ 5 = Detective Division	40 ÷ 5 = Detective Division
35 ÷ 5 = Detective Division	30 ÷ 5 = Detective Division	25 ÷ 5 = Detective Division
20 ÷ 5 = Detective Division	15 ÷ 5 = Detective Division	10 ÷ 5 = Detective Division
5 ÷ 5 = Detective Division	**Build your own** Detective Division	36 ÷ 4 = Detective Division
32 ÷ 4 = Detective Division	28 ÷ 4 = Detective Division	20 ÷ 4 = Detective Division

16 ÷ 4 = Detective Division	12 ÷ 4 = Detective Division	8 ÷ 4 = Detective Division
4 ÷ 4 = Detective Division	**Build your own** Detective Division	27 ÷ 3 = Detective Division
24 ÷ 3 = Detective Division	21 ÷ 3 = Detective Division	18 ÷ 3 = Detective Division
15 ÷ 3 = Detective Division	12 ÷ 3 = Detective Division	9 ÷ 3 = Detective Division
6 ÷ 3 = Detective Division	3 ÷ 3 = Detective Division	**Build your own** Detective Division
18 ÷ 2 = Detective Division	16 ÷ 2 = Detective Division	14 ÷ 2 = Detective Division
12 ÷ 2 = Detective Division	10 ÷ 2 = Detective Division	8 ÷ 2 = Detective Division
6 ÷ 2 = Detective Division	4 ÷ 2 = Detective Division	2 ÷ 2 = Detective Division
Build your own Detective Division	9 ÷ 1 = Detective Division	8 ÷ 1 = Detective Division

7 ÷ 1 = Detective Division	6 ÷ 1 = Detective Division	5 ÷ 1 = Detective Division
4 ÷ 1 = Detective Division	3 ÷ 1 = Detective Division	2 ÷ 1 = Detective Division
1 ÷ 1 = Detective Division	**Build your own** Detective Division	**Build your own** Detective Division

Basic Division Concepts: Greatest Multiples
Developing Procedural Skills and Fluency

AMAZEing Division

(**CCSS:** 3.0A.A.2; 3.0A.A.4; 3.0A.B.5; 3.0A.C.7)

Manipulatives Needed:
- *AMAZEing Division* game board (1 per group)
- 1 die per group of players
- 1 game marker/counter per player
- *AMAZEing* playing cards (1 set per group)

Preparation:
- Copy the *AMAZEing Division* game board onto cardstock (page 141).
- Copy and cut out the *AMAZEing* playing cards (page 142-144). It helps to number them and store them in re-sealable bags so that the sets don't get mixed up.
- Distribute a game marker to each player and the rest of the manipulatives needed to each group.

Game Instructions:
- *AMAZEing* cards are shuffled and placed face down to the right of the *AMAZEing Division* game board.
- Players place their game markers on the *start* space of the maze.
- The first player rolls the die and moves the indicated number of spaces. The number on the space in which they land becomes the divisor.
- The player then turns over the top *AMAZEing* card. This number becomes the dividend.
- Using the number they have landed on and the card they pick, they are to say the greatest multiple that will go into the *AMAZEing* card number.
  - For example, player lands on a 6, picks up a card with the number 46 and states that the largest multiple that will go into 46 is 7.
-
-
- The first player to get to the end of the maze wins.

AMAZEing Division Game board

5 →	6	5	3	7 → End
↑				
2 ←	7	6	9	4 ← 8
				↑
8 →	2	6	5	3 → 2
↑				
6 ←	4	7	9	5 3
				↑
5 →	3	6	7	5 → 8
↑				
9 ← 4		8	5	2 ← 4
				↑
2 →	3	6	9	3 → 7
↑				
8 ← 5		7	4	2 ← 5
Start → 7	8	6	9	3

1	2	3	4
AMAZEing	AMAZEing	AMAZEing	AMAZEing
3	4	5	6
AMAZEing	AMAZEing	AMAZEing	AMAZEing
7	8	9	10
AMAZEing	AMAZEing	AMAZEing	AMAZEing
11	12	13	14
AMAZEing	AMAZEing	AMAZEing	AMAZEing
15	16	17	18
AMAZEing	AMAZEing	AMAZEing	AMAZEing

19 AMAZEing	**20** AMAZEing	**21** AMAZEing	**22** AMAZEing
23 AMAZEing	**24** AMAZEing	**25** AMAZEing	**26** AMAZEing
27 AMAZEing	**28** AMAZEing	**29** AMAZEing	**30** AMAZEing
31 AMAZEing	**32** AMAZEing	**33** AMAZEing	**34** AMAZEing
35 AMAZEing	**36** AMAZEing	**37** AMAZEing	**38** AMAZEing

39 AMAZEing	**40** AMAZEing	**41** AMAZEing	**42** AMAZEing
43 AMAZEing	**44** AMAZEing	**45** AMAZEing	**46** AMAZEing
47 AMAZEing	**48** AMAZEing	**49** AMAZEing	**50** AMAZEing

 Basic Division Concepts: Divisibility
Developing Procedural Skills and Fluency

Divisibility Rules!

(CCSS: **(CCSS**: 3.OA.A.2; 3.OA.5; 3.OA.6; 4.OA,B,4 ;4.NBT.A.1)

Manipulatives Needed:
- *Divisibility Rules* game board (1 per group of 2-3 students)
- *Divisibility* spinner (1 per group of 2-3 students)
- *Divisibility Rules* tiles (1 per group of 2-3 students) or blank tiles with numbers 0-9 [7 each] written on them.
- *Calculator* (optional for keeping score)

Preparation:
- Copy the game board onto cardstock, cut apart and tape the 4 pieces together so that the shaded border is around the outside of the *Divisibility Rules* game board (page 150-153).
- Copy the *Divisibility Rules* tiles onto cardstock and cut apart (page 46-48).
- Assemble the divisibility spinner (page 149) using assembly instructions on page 16.

Game Instructions:
- *Divisibility Rules* is played similarly to the traditional game of Scrabble.
- After all the tiles are mixed up and placed face down, players select 7 tiles.
- First player spins the spinner and uses however many tiles they choose to build a number that is divisible by whatever number is spun. They may place tiles on board in a horizontal or vertical pattern.
- Player receives points for each tile used (e.g., if a 2 is used that counts as 2 points) plus the bonus points indicated on the spinner.
- If a tile is placed over a star, the total points of that tile are doubled.
- If tiles intersect they must work for whatever new number is created.
- Player selects new tiles to replace the number they used.
- The next player must overlap at least 1 number from previous number played.
- The first player to earn 100 points wins.
- Calculators can be used to keep score and/or check the divisibility of numbers played on board.

 **3,261** is divisible by 3. 18 pts. scored on play + 2 bonus pts = 20 pts.

0	0	0	0
0	0	0	1
1	1	1	1
1	1	2	2
2	2	2	2
2	3	3	3

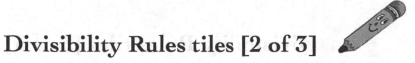

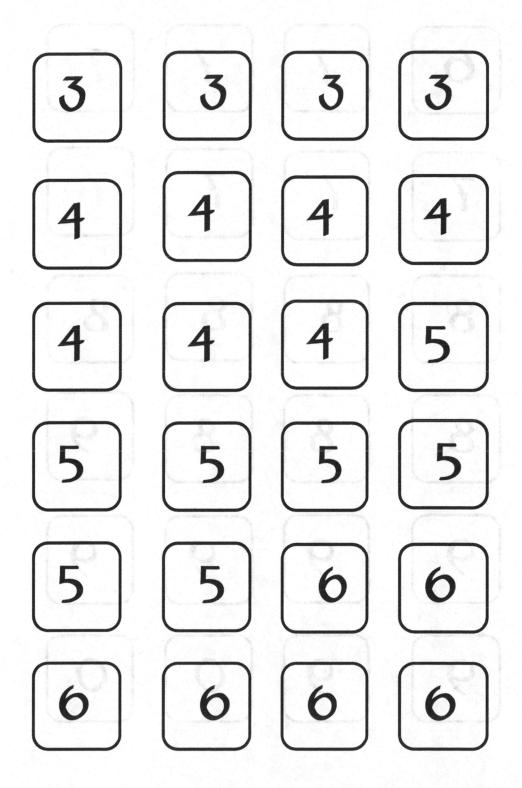

3	3	3	3
4	4	4	4
4	4	4	5
5	5	5	5
5	5	6	6
6	6	6	6

6	7	7	7
7	7	7	7
8	8	8	8
8	8	8	9
9	9	9	9
9	9	0	0

Divisibility Rules! spinner

9
Sum of digits is divisible by 9

2
1 point
Last digit is even (0,2,4,6,8)

8
Last 3 digits are divisible by 8

3
2 points
Sum of the digits are a multiple of 3

Double last digit. Subtract from remaining number. Repeat to see if divisible by 7

Last 2 digits are a multiple of 4

7
5 points

4
3 points
Last digit is a 0 or 5

6
4 points

Passes tests for 2 and 3

5
1 point

Multi-Digit Division

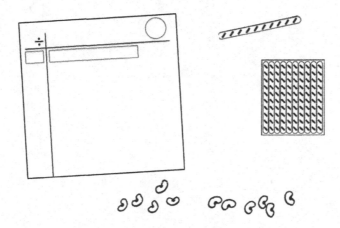

Acquiring conceptual understanding with multi-digit division allows students to implement basic division facts and gain fluency with strategies for complex division problems.

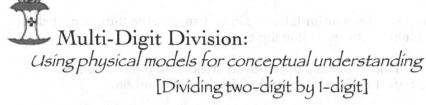

Multi-Digit Division:
Using physical models for conceptual understanding
[Dividing two-digit by 1-digit]

Division is for the Beans! [A]

(CCSS:4.0A.A.3; 4.NBT.B.6; 5.NBT.B.6; 6.NS.2)

Manipulatives Needed:

- Student beanstick sets (see below) or base-ten blocks
- *Division is for the Beans!* work mat (1 per student or group)
- *Division is for the Beans*! [A] Worksheet (1 per student or group)
- *Division is for the Beans*! Blank Worksheet (1 per student or group)

Preparation:

- Copy *Division is for the Beans!* workmat and laminate or put in plastic sleeves (page 161).
- Copy *Division is for the Beans![A]* Problem Record Sheet for each student/group (page 162).
- Students and/or teacher can use the blank *Division is for the Beans!* blank worksheet (page 163) to create a variety of problems to solve.
- Construct beansticks and hundred-bean flats as follows (or use base-ten blocks).
    - **Materials Needed**:
        - Small red beans
        - Popsicle sticks
        - White glue
        - Heavy cardboard
        - Cardstock
        - Paper cutter
    - **Ten-beanstick construction:** Apply white glue liberally to a popsicle stick and place 10 beans evenly in the glue. Follow with a final coat of glue over the tops of the bean and in between the beans to prevent chipping. Let the glue dry for 24 hours before using.
    - **Hundred-Bean Flat:** Glue ten beansticks side by side onto heavy cardboard. Cut out around the perimeter of the beansticks. Copy the hundred-bean flat drawings pattern on page 160 to use for additional flats.
    - **Student beanstick set:** A set for one student consists of 1-2 real hundred-bean flats, 9 drawings of flats, 15 ten-beansticks, and 20 loose beans. Store the sets in re-closable bags.

Procedure:

- Demonstrate how the manipulatives can be use to solve 2-digit by 1-digit division problems with the following example:

    66 ÷ 5   Take 66 beans (6 ten-bean sticks and 6 loose beans)
    Write 5 in the divisor box and 66 in the dividend box

  Use the array method (creating a rectangle of even beans) with 5 rows of an even amount of the 66 beans in the workspace of the *Division is for the Beans!* work mat. If there are beans left that cannot be shared evenly, those become the remainder.

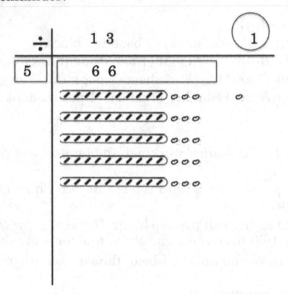

- Demonstrate how students show their work on the *Division is for the Beans!* Record Sheet by drawing lines and dots to represent the beansticks and beans as   illustrated:

Special Note: After students have worked with the manipulatives, you will find they transition quickly to solving problems with drawings. Building a solid foundation will also lead to an easy and visual transition to the thought process behind long division.

Answers to *Division is for the Beans!* [A]:
1. **4r1**
2. **23 r 1**
3. **17 r 1**
4. **8 r 2**
5. **17**
6. **19**
7. **6 r 3**
8. **28 r 1**
9. **43**
10. **13**
11. **9 r 1**
12. **18 r 3**
Check student's work for accuracy.

Hundred-Bean Flat Pattern

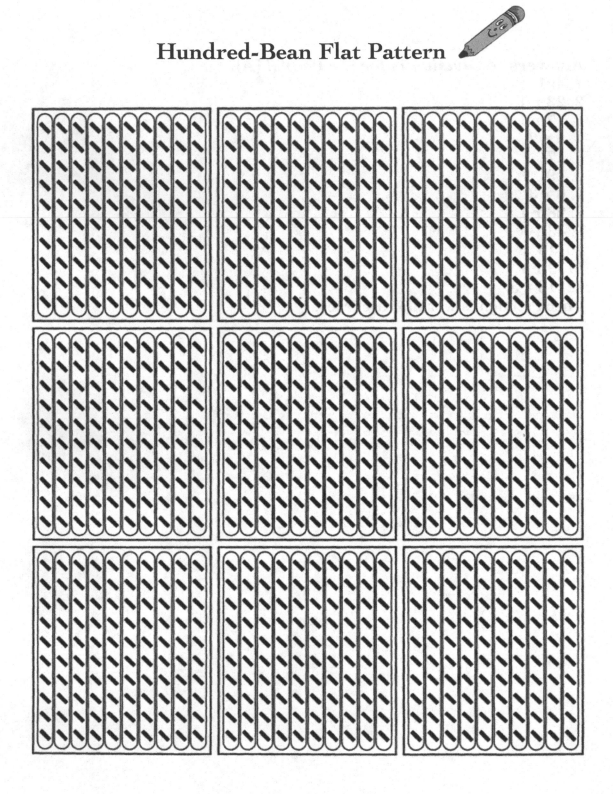

Division is for the Beans! [work mat]

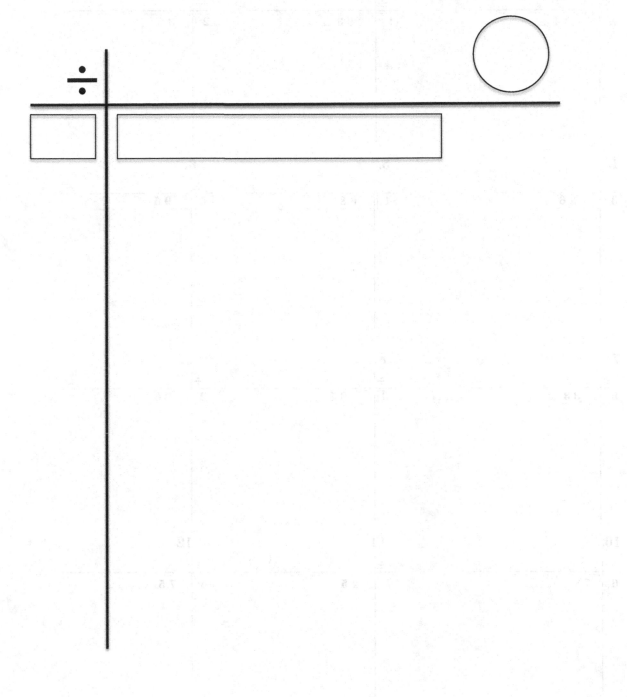

Division is for the Beans! [A] worksheet

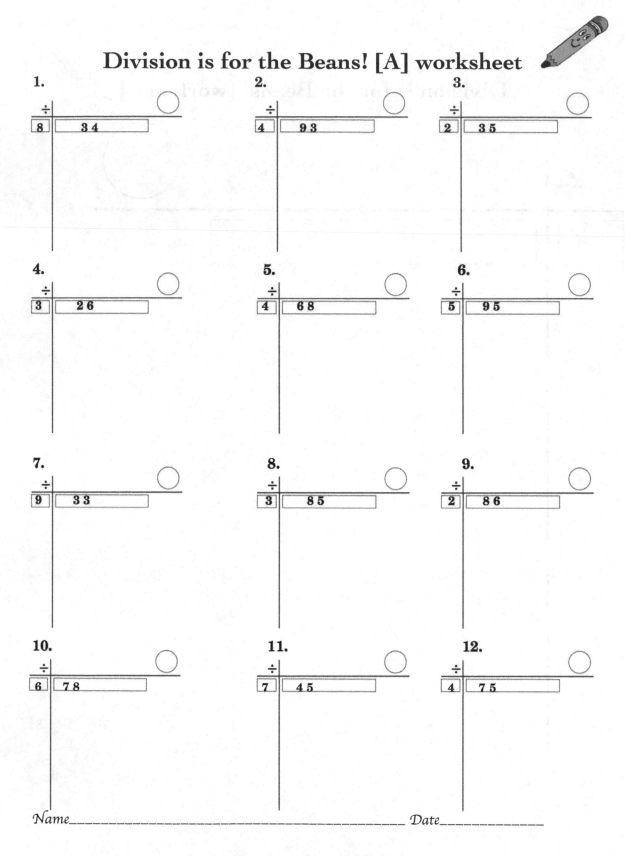

1.

$8 \div \overline{ 3\,4 }$ ◯

2.

$4 \div \overline{ 9\,3 }$ ◯

3.

$2 \div \overline{ 3\,5 }$ ◯

4.

$3 \div \overline{ 2\,6 }$ ◯

5.

$4 \div \overline{ 6\,8 }$ ◯

6.

$5 \div \overline{ 9\,5 }$ ◯

7.

$9 \div \overline{ 3\,3 }$ ◯

8.

$3 \div \overline{ 8\,5 }$ ◯

9.

$2 \div \overline{ 8\,6 }$ ◯

10.

$6 \div \overline{ 7\,8 }$ ◯

11.

$7 \div \overline{ 4\,5 }$ ◯

12.

$4 \div \overline{ 7\,5 }$ ◯

Name_____ Date_____

Division is for the Beans! [blank worksheet]

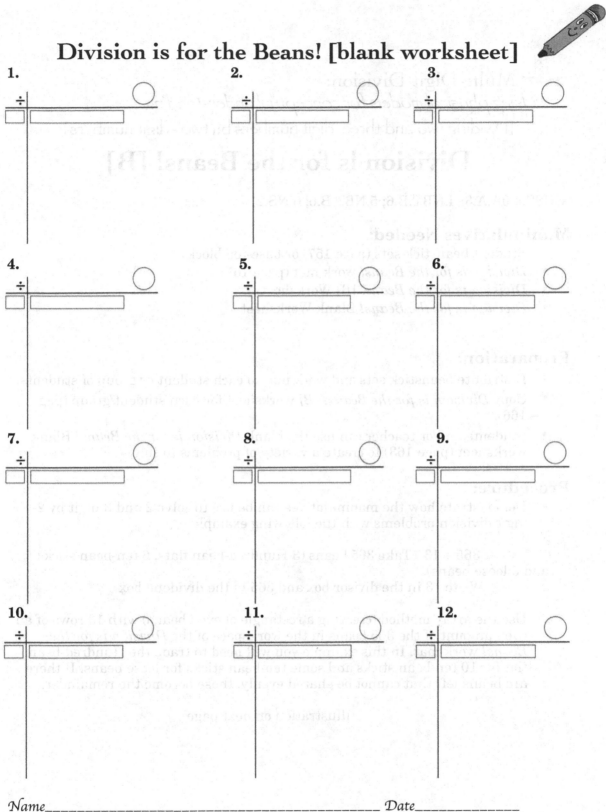

1.

2.

3.

4.

5.

6.

7.

8.

9.

10.

11.

12.

Name_____ Date_____

Multi-Digit Division:
Using physical models for conceptual understanding
[Dividing two and three-digit numbers by two -digit numbers]

Division is for the Beans! [B]

(CCSS:4.0A.A.3; 4.NBT.B.6; 5.NBT.B.6; 6.NS.2)

Manipulatives Needed:
- Student beanstick sets (page 157) or base-ten blocks
- *Division is for the Beans!* work mat (page 161)
- *Division is for the Beans!* [B] Worksheet
- *Division is for the Beans!* Blank Worksheet

Preparation:
- Distribute beanstick sets and work mat to each student or group of students.
- Copy *Division is for the Beans! [B]* worksheet for each student/group (page 166).
- Students and/or teacher can use the blank *Division is for the Beans!* Blank worksheet (page 163) to create a variety of problems to solve.

Procedure:
- Demonstrate how the manipulatives can be use to solve 2 and 3-digit by 2-digit division problems with the following example:

     $365 \div 13$ Take 365 beans (3 Hundred-bean flats, 6 ten-bean sticks and 5 loose beans).
     Write 13 in the divisor box and 365 in the dividend box

Use the array method (creating a rectangle of even beans) with 13 rows of an even amount of the 365 beans in the workspace of the *Division is for the Beans!* work mat. In this example you will need to trade the Hundred-bean flat for 10 ten-bean sticks and some ten-bean sticks for loose beans. If there are beans left that cannot be shared evenly, those become the remainder.

(illustration on next page)

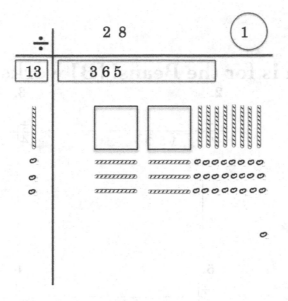

- Demonstrate how students show their work on the *Division is for the Beans!* worksheet by drawing squares, lines and dots to represent the hundred-flat, beanstick and beans.

Special Note: After students have worked with the manipulatives, you will find they transition quickly to solving problems with drawings. Building a solid foundation will also lead to an easy and visual transition to the thought process behind long division.

Answers to *Division is for the Beans!* [B]:
1. **9**
2. **4 r 2**
3. **7 r 1**
4. **3**
5. **8 r 37**
6. **1 r 2**
7. **2 r 8**
8. **1 r 2**
9. **6**
10. **13 r 7**
11. **21 r 18**
12. **44 r 15**

Check student's work for accuracy.

Division is for the Beans! [B] worksheet

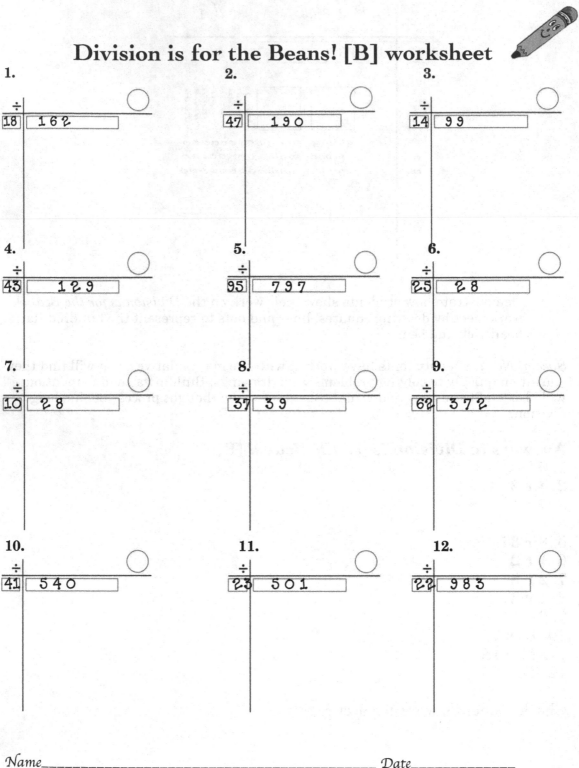

1.

$18 \div 162$ ◯

2.

$47 \div 190$ ◯

3.

$14 \div 99$ ◯

4.

$43 \div 129$ ◯

5.

$95 \div 797$ ◯

6.

$25 \div 28$ ◯

7.

$10 \div 28$ ◯

8.

$37 \div 39$ ◯

9.

$62 \div 372$ ◯

10.

$41 \div 540$ ◯

11.

$23 \div 501$ ◯

12.

$22 \div 983$ ◯

Name_____ Date_____

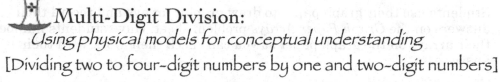

Multi-Digit Division:
Using physical models for conceptual understanding
[Dividing two to four-digit numbers by one and two-digit numbers]

Extending Division with Graph Paper Arrays

(**CCSS**: 5.NBT.6; 6.NS.2)

Manipulatives Needed:
- Multi-colored small grid graph paper
- *Division Graph Paper Arrays* worksheet

Preparation:
- Make 4 copies of multi-colored graph paper per student (page 170).
- Copy the *Division Graph Paper Arrays* worksheet for each student (page 169).

Procedure:
- Demonstrate how to use the graph paper to draw and solve division problems as illustrated below:

$$276 \div 23 = 12$$

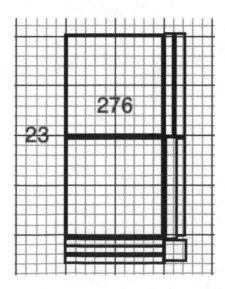

- Students use their graph paper to draw graphic arrays and record their answers on the *Graph Paper Arrays* problem sheet. Have students number their arrays on the graph paper so you can check their work with their answers.

Answers:

1. 9
2. 3
3. 2
4. 9
5. 3
6. 6
7. 4
8. 8
9. 5
10. 4
11. 4
12. 7

Division Graph Paper Arrays

[1] 35)315   [2] 68)204   [3] 72)144   [4] 51)459

[5] 23)69   [6] 81)486   [7] 95)380   [8] 50)400

[9] 23)115   [10] 33)132   [11] 63)252   [12] 84)588

Name_____ Date_____

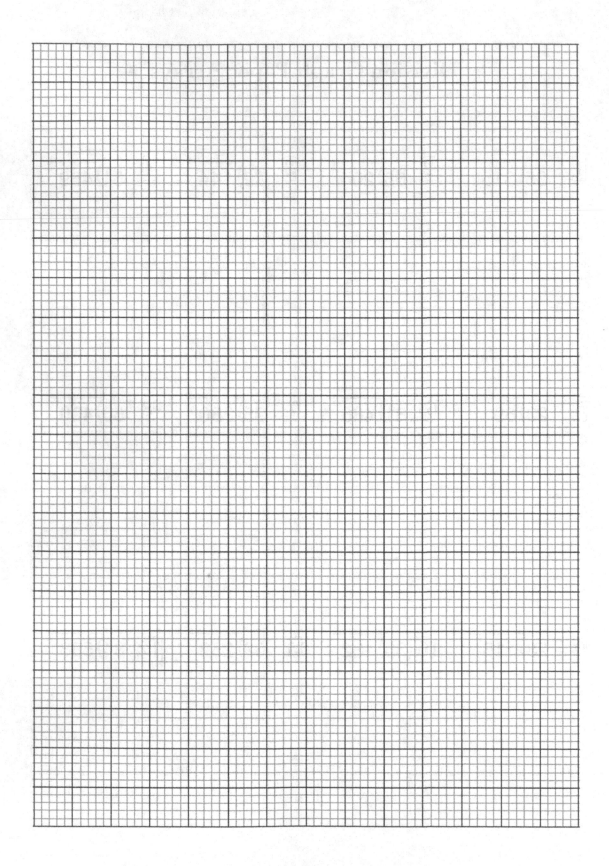

 Multi-Digit Division

Developing Procedural Skills and Fluency

Mastering and Managing Division with the Help of Monkeys!

(CCSS: 5.NBT.6; 6.NS.2)

Manipulatives Needed:
- 1 cm graph paper (page 14)
- *Division with the Help of Monkeys* worksheet (1 per student)

Preparation:
- Copy *Division with the Help of Monkeys* (page 174).
- Copy or purchase graph paper for additional practice problems.

Procedure:
- Teach the process of division with the following acronym and graph paper to manage the correct placement of the digits:
    Do Monkeys Steal Bananas? Restock or Run
    - o **Do** (for divide)
    - o **Monkeys** (for multiply)
    - o **Steal** (for subtract)
    - o **Bananas** (for bring down)
    - o **Repeat** (if you can still divide) or **Restock** (for remainder)

Example:

		3	5	5	r3	Thought Process
2	7	1	3			**Do - Divide** – How many times does 2 go into 7? 3
-	6					**Monkeys** – Multiply 2x3=6
	1	1				**Steal** – Subtract 7-6=1
-	1	0				**Bananas** – **B**ring down next digit =1 repeat….
		1	3			**Do** – **D**ivide – How many times does 2 go into 11? 5
	-	1	0			**Monkeys** – Multiply 5x2=10
			3			**Steal** – **S**ubtract 11-10=1
						Bananas – Bring down next digit – 3 repeat….
						Do – **D**ivide – How many times does 2 go into 13? 5
						Monkeys – **M**ultiply 2 x 5 = 10
						Steal – **S**ubtract 13-10 = 3 **Restock** as 3 is the **remainder** and this monkey will steal again ☺

- I have found it effective to teach the process with problems that require repetition of the process several times within a problem. Use *Division with the Help of Monkeys* worksheet as a starter set of problems to reinforce the process.
- If a divisor is larger than one of the digits, teach students to place an x in that spot combine with next digit. For example:

	x		
7	2 9		

- If the divisor has more than one digit then instruct students to underline the same number in the dividend to start. For example:

		x	x	
2	3	5	2	5

Do Monkeys Steal Bananas? Yes they do – so be prepared for them to repeat or you restock and have fun with division!

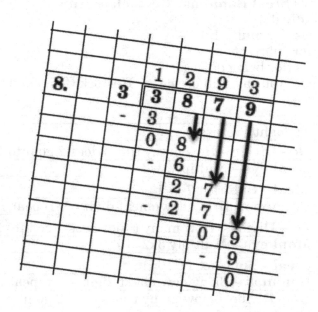

Division with the Help of Monkeys
Answer sheet

1.   quotient: 1 6 7 r1
```
      1 6 7 r1
 5 ) 8 3 6
     5
   - 3 3
     3 0
       3 6
     - 3 5
         1
```

2.   quotient: x 4 8 r2
```
      x 4 8 r2
 8 ) 3 8 6
   - 3 2
       6 6
     - 6 4
         2
```

3.   quotient: x 7 7 r3
```
      x 7 7 r3
 6 ) 4 6 5
   - 4 2
       4 5
     - 4 2
         3
```

4.   quotient: x 8 9 r1
```
      x 8 9 r1
 4 ) 3 5 7
   - 3 2
       3 7
     - 3 6
         1
```

5.   quotient: x 7 5
```
      x 7 5
 7 ) 5 2 5
   - 4 9
       3 5
     - 3 5
         0
```

6.   quotient: 4 5 7 r1
```
      4 5 7 r1
 2 ) 9 1 5
   - 8
       1 1
     - 1 0
         1 5
       - 1 4
           1
```

7.   quotient: x 3 2 9
```
      x 3 2 9
 8 ) 2 6 3 2
   - 2 4
       0 2 3
       - 1 6
           7 2
         - 7 2
             0
```

8.   quotient: 1 2 9 3
```
      1 2 9 3
 3 ) 3 8 7 9
   - 3
       0 8
       - 6
           2 7
         - 2 7
             0 9
           - 9
               0
```

9.   quotient: x 2 7
```
      x 2 7
 9 ) 2 4 3
   - 1 8
       6 3
     - 6 3
         0
```

10   quotient: 1 1 4 8 2
```
        1 1 4 8 2
 5 ) 5 7 4 1 0
   - 5
     0 7
   - 5
     2 4
   - 2 0
       4 1
     - 4 0
         1 0
       - 1 0
           0
```

   quotient: 1 2 3 4 1
```
        1 2 3 4 1
 6 ) 7 4 0 4 6
   - 6
     1 4
   - 1 2
       2 0
     - 1 8
         2 4
       - 2 4
           0 6
         - 6
             0
```

Division with the Help of Monkeys

1.		5	8	3	6		2.		8	3	8	6		3.		6	4	6	5

4.		4	3	5	7		5.		7	5	2	5		6.		2	9	1	5

7.		8	2	6	3	2		8.		3	3	8	7	9		9.		9	2	4	3

10		5	5	7	4	1	0		11		6	7	4	0	4	6

Name_____ Date_____

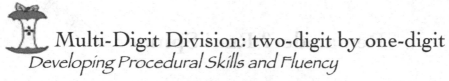 Multi-Digit Division: two-digit by one-digit
Developing Procedural Skills and Fluency

Remainder? Run!

(**CCSS**: 4.MBT.6; 5.MB.5; 6.NS.2)

Manipulatives Needed:
- *Remainder? Run!* spinner per group of 2-4 players
- *Remainder? Run!* game board per group of 2-4 players
- 1 game marker for each player

Preparation:
- Copy the *Remainder? Run!* game board onto cardstock for durability (page 177).
- Assemble the *Remainder? Run!* spinner (page 176).
- Distribute game manipulatives to player and group.

Game Instructions:
- Each player spins the spinner to see who goes first. Highest number goes first and play continues clockwise. Markers are placed on the number above *Start*.
- First player spins. This number is the divisor for the dividend (number on space where marker is).
  - *Example*: 8 is spun. Marker is on 54. How many 8's in 54? There are 6. What is the remainder? 6. Player moves 6 spaces.
  - *Example*: 6 is spun. Marker is on 36. Since there is not a remainder, player doesn't move on that turn and tries again on their next turn.
- First player to reach *Stop* wins.

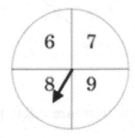

Remainder? Run! Game board

54 Start	18	35	24	28	20	(37)	58
							27
							19

Remainder? Run! Spinner

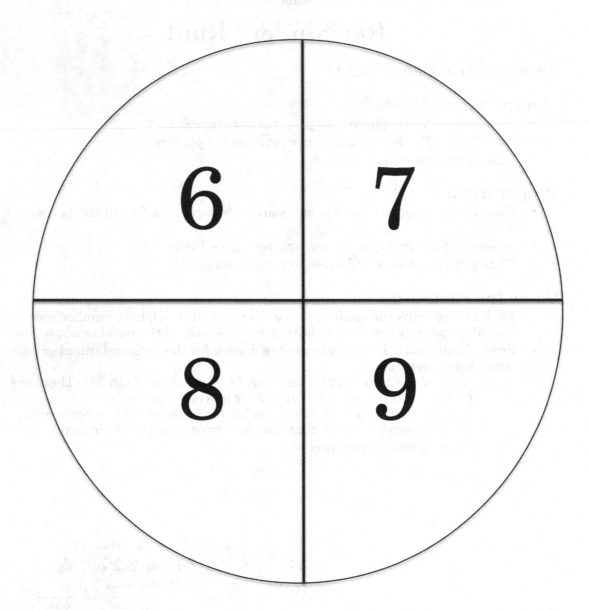

To assemble, duplicate the pattern onto cardstock. Use a brad to secure safety pin in center of spinner.

Remainder? Run! Game board

54 Start	18	35	24	28	20	37	58
							27
							19
22	55	36	30	29	45	52	44
50							
31	47	21	56	39	57	43	46
							53
41	48	51	38	40	34	26	32
33	25	59	24	49	72	64	Stop

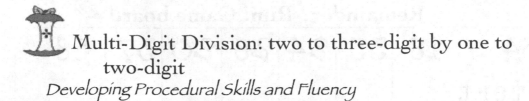 Multi-Digit Division: two to three-digit by one to two-digit

Developing Procedural Skills and Fluency

Headed Home!

(**CCSS**: 4,NBT.6; 5.NBT.6; 6.NS.2)

Manipulatives Needed:
- *Headed Home!* Game board for each group of 2-4 students
- *Headed Home!* Playing cards for each group of 2-4 students
- 1 game marker for each player

Preparation:
- Copy *Headed Home!* game board onto cardstock for durability (page 181).
- Copy *Headed Home!* cards onto cardstock and cut apart and store in re-closable bags (page 179-180).
- Distribute game manipulatives to each player and group.

Game Instructions:
- Shuffle the *Headed Home* cards and place them face down next to the gameboard.
- Players place their game markers on the *start* space. First player turns over the top card in the stack and uses that card as the divisor. The start number of 131 is the dividend. All players write and solve the division problem on a piece of paper. Answers are compared and the player who drew the card moves the amount of spaces in the remainder if they are correct.
- If no one reaches home after all the cards have been used, shuffle the cards and continue playing.
- Players may occupy the same space.
- First player to reach home wins.

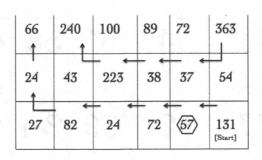

Example: 131 ÷ 5 = 26 r 1 (so player moves 1 space)

Headed Home! [Cards 1 of 2]

Headed Home!	Headed Home!	Headed Home!
1	**2**	**3**
Headed Home!	Headed Home!	Headed Home!
4	**5**	**6**
Headed Home!	Headed Home!	Headed Home!
7	**8**	**9**
Headed Home!	Headed Home!	Headed Home!
10	**11**	**12**

Headed Home!	Headed Home!	Headed Home!
13	**14**	**15**
Headed Home!	Headed Home!	Headed Home!
16	**17**	**18**
Headed Home!	Headed Home!	Headed Home!
19	**20**	**21**
Headed Home!	Headed Home!	Headed Home!
22	**23**	**24**

Headed Home! [Game board]

311	25	115	49	271	29
216	96	43	31	109	464
65	78	HOME		793	903
66	240	100	89	72	363
24	43	223	38	37	54
27	82	24	72	57	131 [Start]

Multi-Digit Division: three to four- digit by one to two- digit

Developing Procedural Skills and Fluency

Tally Oh!

(**CCSS**: 4.NBT.6; 5.NBT.6; 6.NS.2)

Manipulatives Needed:
- *Tally Oh!* Record sheet per player
- 6 *Octahedron* dice per group of 2-5 players
- Graph paper to show work

Preparation:
- Copy *Tally oh!* record sheet for each player.
- Copy the *Octahedron die pattern* (page 100) onto cardstock and assemble according to the directions on the pattern.
- Distribute game manipulatives to each player and group.

Game Instructions:
- Divide the students into teams of 2-6 players.
- Before the start of each round, group decides on how many dice they will each roll and what type of problem they are creating. For example, they can decide to roll 4 dice and create 3 digit by 1 digit problems.
- Each player will roll the agreed upon number of dice and create a problem with the numbers rolled.
- Problem is recorded on the *Tally Oh!* record sheet and the remainder becomes their tally points for that round.
- Play continues for up to 10 rounds. The player with the fewest tally points is the winner.

Tally Oh! [Record sheet]

Round	Number of dice	Type	Division problem	Answer	Tally points
example	5	4 by 2	345 ÷ 12	28 r9	9
1	4	3 by 1	732 ÷ 5	146 r 2	2
2					

Tally Oh! [Record sheet]

Round	Number of dice	Type	Division problem	Answer	Tally points
example	5	4 by 2	345 ÷ 12	28 r9	9
1					
2					
3					
4					
5					
6					
7					
8					
9					
10					

Name_____ Date_____

Multi-Digit Division: three-digit by one-digit
Developing Procedural Skills and Fluency

Get Rich Quick!

(**CCSS**: 4.NBT.6; 5.NBT6; 6.NS.2)

Manipulatives Needed:
- 4 *Octahedron dice* (2 sets per class)
- *Get Rich Quick!* Money (1 set per team)
- Graph paper (page 14)

Preparation:
- Copy and cut apart enough *Get Rich Quick!* money for two sets (page 184-188)
- Assemble 4 octahedron dice (see page 100 for octahedron die pattern and assembly instructions).
- Set up two tables on opposite sides of the room with a set of money and graph paper and pencils.

Game Instructions:
- Divide the class into two teams. Put teams on each side of the room behind the tables.
- Roll 4 octahedron dice to create a three-digit by one-digit division problems for teams to solve with the money.
    > Example: A 345 and a 4 are rolled. Tell teams they have $345 dollars (3 one-hundred dollar bills, 4 ten-dollar bills, and 5 one-dollar bills). They need to buy presents for four of their friends. What is the most they can spend on each friend if they give a present of the same value to each? Separate the money into 4 piles. They may need to exchange larger bills for smaller ones.
- Once the problem has been announced to the teams, 2 students are to come to the tables and solve the problem together, recording their results on the graph paper.
- The team that correctly finishes first scores a point.
- Continue activity until all the students in each team have had a turn.

Variation: This activity can be set up in groups of 2-4 students, as a learning center or as an individual activity to practice multi-digit division.

Get Rich Quick! [money]
(copy and cut out 10 sheets per page per set)

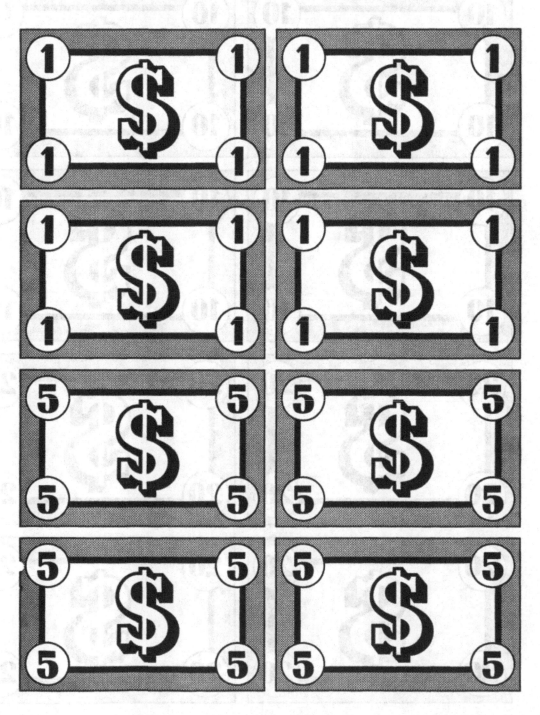

Get Rich Quick! [money]

(copy and cut out 10 sheets per page per set)

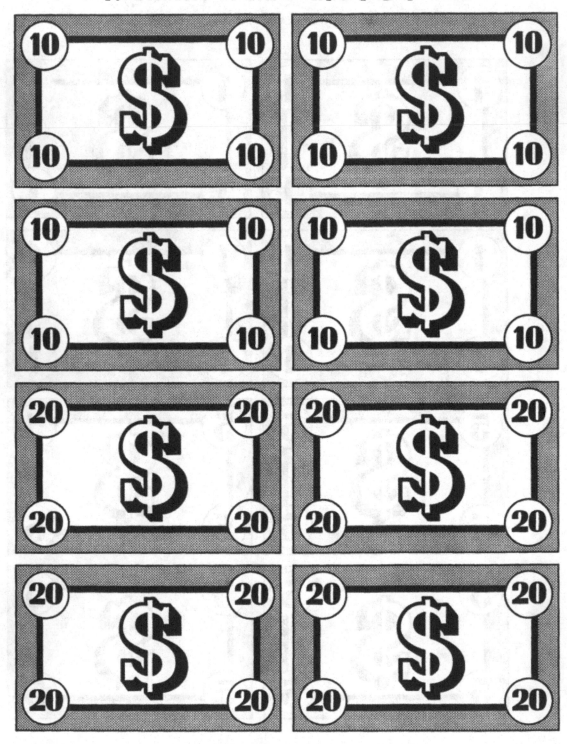

Get Rich Quick! [money]

(copy and cut out 10 sheets per page per set)

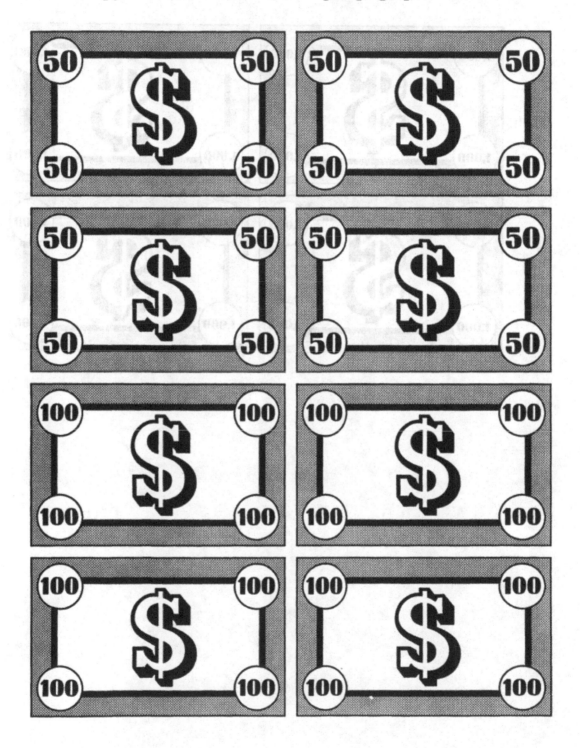

Get Rich Quick! [money]
(copy and cut out 10 sheets per set)

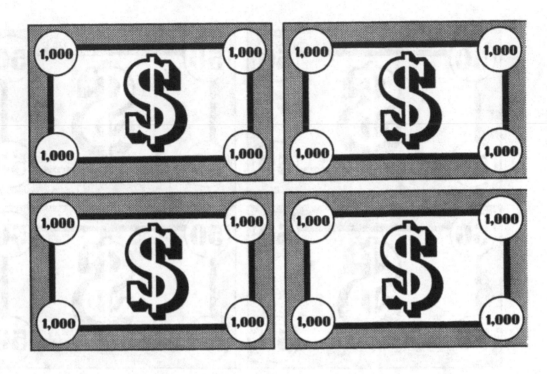

Reinforcing Multiplication and Division

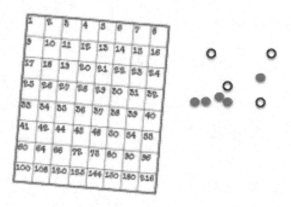

Rummy	Rummy	Rummy	Rummy
24÷3=8	24÷8=3	3x8=24	8x3=24
Style	Style	Style	Style

Reinforcing Multiplication and Division
Developing Procedural Skills and Fluency

Block It!

(CCSS: 3.0A.A.3; 3.0A.C.7;4.NBT.B.5;4.NBT.B.5; 4.NBT.B.6; 5.NBT.B.5;.5.NBT.B.6;6.NS.2)

Manipulatives Needed:
- *Block It!* game board for each team of 2 students
- *Block It!* dice for each team of 2 students
- 32 counters for each player (two-sided counters work great)
- Graph paper to show work (page 14)

Preparation:
- Copy *Block It!* game board onto cardstock for durability for each team (page 192).
- Copy onto cardstock and assemble the 3 *Block It!* dice (page 193-195).
- Distribute game manipulatives to each team.

Game Instructions:
- Divide class into teams of 2 players.
- Each player rolls the three *Block It!* dice. Using any combination of addition, subtraction, multiplication or division, they use the three dice to try and create a number on the board showing their work on the graph paper.
- Player chooses their color side (red or yellow) and place their counter in the square that is the answer to their problem they created. They cannot place a counter on a square that is already occupied.
- The object of the game is to place 4 markers in adjacent squares in a row, column, diagonal or in a two by two square. Player tries to block their opponent's arrangements.

1	2	3	4	5	6	7	8
9	10	11	12	13	14	15	16
17	18	19	20	21	22	23	24
25	26	27	28	29	30	31	32
33	34	35	36	37	38	39	40
41	42	44	45	48	50	54	55
60	64	66	72	75	80	90	96
100	108	120	125	144	150	180	216

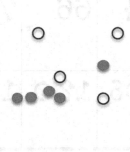

Block It! [game board]

1	2	3	4	5	6	7	8
9	10	11	12	13	14	15	16
17	18	19	20	21	22	23	24
25	26	27	28	29	30	31	32
33	34	35	36	37	38	39	40
41	42	44	45	48	50	54	55
60	64	66	72	75	80	90	96
100	108	120	125	144	150	180	216

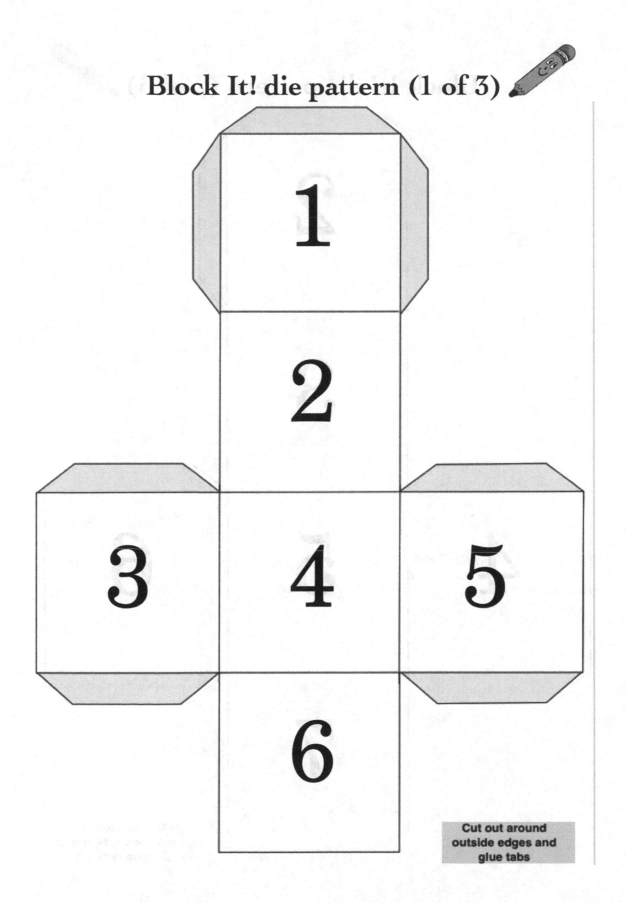

Cut out around
outside edges and
glue tabs

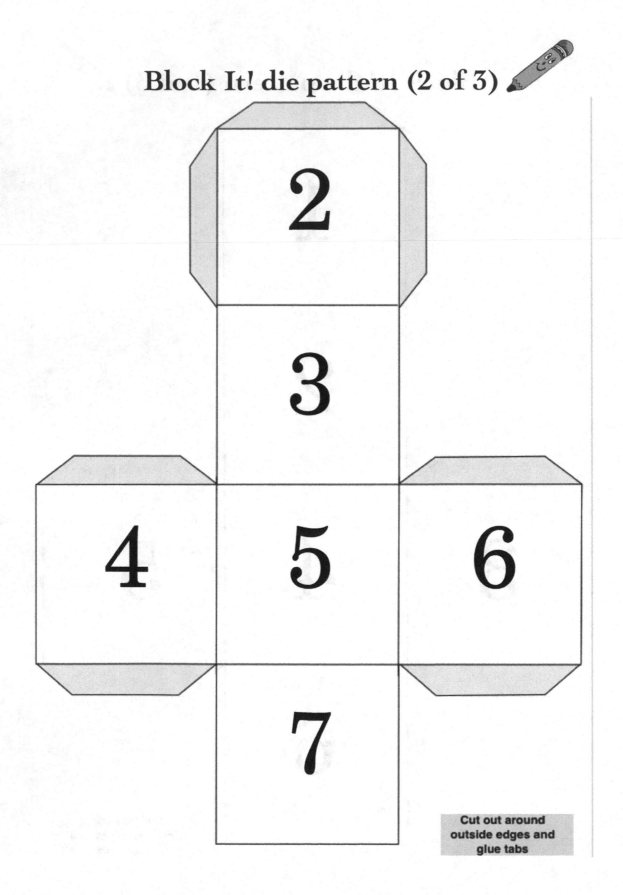

Cut out around outside edges and glue tabs

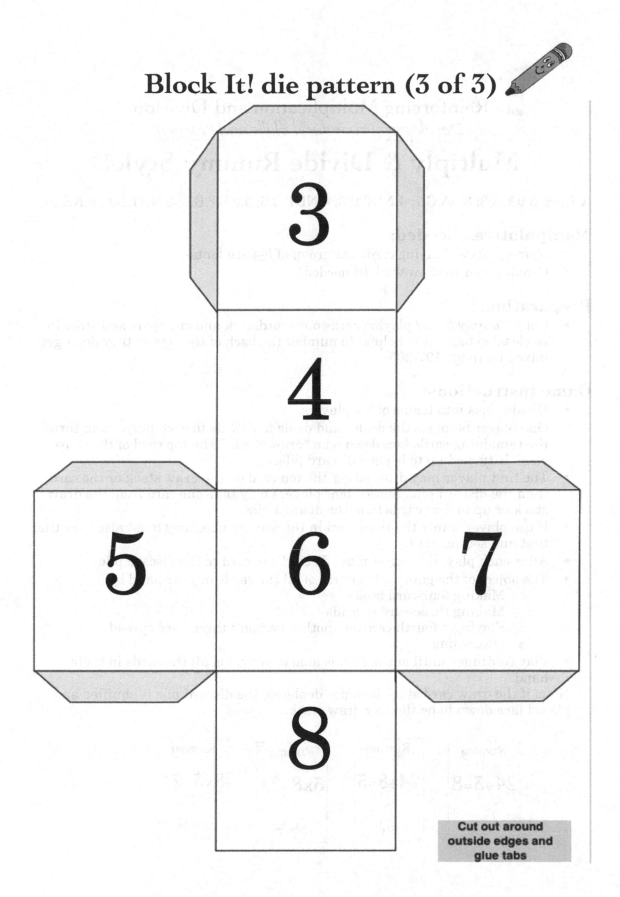

3

4

5   6   7

Cut out around outside edges and glue tabs

8

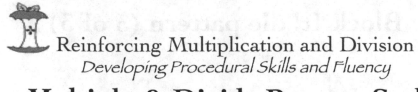

Reinforcing Multiplication and Division
Developing Procedural Skills and Fluency

Multiply & Divide Rummy Style!

(**CCSS**: 3.OA.A.3; 3.OA.C.7; 4.NBT.B.5; 4.NBT.B.6; 5.NBT.B.5; 5.NBT.B.6; 6.NS.2)

Manipulatives Needed:
- *Rummy Style!* Playing cards per group of 3-4 students
- Graph paper to show work (if needed)

Preparation:
- Copy *Rummy Style!* playing cards onto cardstock and cut apart and store in re-closable bags. It is helpful to number the back of the sets so they don't get mixed up (page 197-205).

Game Instructions:
- Divide class into teams of 3-4 players.
- One player becomes the dealer and deals five cards to each player and turns the remaining cards face down on a "draw stack." The top card of the draw stack is turned up to begin a discard pile.
- The first player may draw either the top card on the draw stack or the card from the discard pile. Subsequent players may take one card from the draw stack or up to four cards from the discard pile.
- If the player wants the third card in the discard pile, they must also take the first and second card.
- After each play, the player must discard one card to the discard pile.
- The object of the game is to get rid of all the cards in your hand by:
  - Making four-card books
  - Making three-card spreads
  - Playing a fourth card on another person's three-card spread
  - Discarding
- Play continues until one person is able to get rid of all the cards in their hand.

Note: if the draw card stack becomes depleted, the discard pile is shuffled and placed face down to be the new draw stack.

Rummy	Rummy	Rummy	Rummy
$24 \div 3 = 8$	$24 \div 8 = 3$	$3 \times 8 = 24$	$8 \times 3 = 24$
Style	Style	Style	Style

Rummy Style [playing cards] (1 of 9)

Rummy	Rummy	Rummy	Rummy
$9 \times 8 = 72$	$8 \times 9 = 72$	$72 \div 9 = 8$	$72 \div 8 = 9$
Style	Style	Style	Style
Rummy	Rummy	Rummy	Rummy
$8 \times 7 = 56$	$7 \times 8 = 56$	$56 \div 8 = 7$	$56 \div 7 = 8$
Style	Style	Style	Style
Rummy	Rummy	Rummy	Rummy
$8 \times 6 = 48$	$6 \times 8 = 48$	$48 \div 8 = 6$	$48 \div 6 = 8$
Style	Style	Style	Style

Rummy	Rummy	Rummy	Rummy
8x6=48	6x8=48	48÷8=6	48÷6=8
Style	Style	Style	Style
Rummy	Rummy	Rummy	Rummy
8x5=40	5x8=40	40÷8=5	40÷5=8
Style	Style	Style	Style
Rummy	Rummy	Rummy	Rummy
8x4=32	4x8=32	32÷8=4	32÷4=8
Style	Style	Style	Style

Rummy	Rummy	Rummy	Rummy
8x3=24	3x8=24	24÷8=3	24÷3=8
Style	Style	Style	Style
Rummy	Rummy	Rummy	Rummy
8x2=16	2x8=16	16÷8=2	16÷2=8
Style	Style	Style	Style
Rummy	Rummy	Rummy	Rummy
7x9=63	9x7=63	63÷9=7	63÷7=9
Style	Style	Style	Style

Rummy	Rummy	Rummy	Rummy
$7 \times 6 = 42$	$6 \times 7 = 42$	$42 \div 7 = 6$	$42 \div 6 = 7$
Style	Style	Style	Style
Rummy	Rummy	Rummy	Rummy
$7 \times 5 = 35$	$5 \times 7 = 35$	$35 \div 7 = 5$	$35 \div 5 = 7$
Style	Style	Style	Style
Rummy	Rummy	Rummy	Rummy
$7 \times 4 = 28$	$4 \times 7 = 28$	$28 \div 7 = 4$	$28 \div 4 = 7$
Style	Style	Style	Style

Rummy	Rummy	Rummy	Rummy
$7 \times 3 = 21$	$3 \times 7 = 21$	$21 \div 7 = 3$	$21 \div 3 = 7$
Style	Style	Style	Style
Rummy	Rummy	Rummy	Rummy
$6 \times 9 = 54$	$9 \times 6 = 54$	$54 \div 9 = 6$	$54 \div 6 = 9$
Style	Style	Style	Style
Rummy	Rummy	Rummy	Rummy
$6 \times 5 = 30$	$5 \times 6 = 30$	$30 \div 5 = 6$	$30 \div 6 = 5$
Style	Style	Style	Style

Rummy	Rummy	Rummy	Rummy
$6 \times 4 = 24$	$4 \times 6 = 24$	$24 \div 4 = 6$	$24 \div 6 = 4$
Style	Style	Style	Style
Rummy	Rummy	Rummy	Rummy
$6 \times 3 = 18$	$3 \times 6 = 18$	$18 \div 6 = 3$	$18 \div 3 = 6$
Style	Style	Style	Style
Rummy	Rummy	Rummy	Rummy
$6 \times 2 = 12$	$2 \times 6 = 12$	$12 \div 6 = 2$	$12 \div 2 = 6$
Style	Style	Style	Style

Rummy	Rummy	Rummy	Rummy
5x9=45	9x5=45	45÷9=5	45÷5=9
Style	Style	Style	Style
Rummy	Rummy	Rummy	Rummy
5x4=20	4x5=20	20÷4=5	20÷5=4
Style	Style	Style	Style
Rummy	Rummy	Rummy	Rummy
5x3=15	3x5=15	15÷3=5	15÷5=3
Style	Style	Style	Style

Rummy	Rummy	Rummy	Rummy
5x2=10	2x5=10	10÷5=2	10÷2=5
Style	Style	Style	Style
Rummy	Rummy	Rummy	Rummy
4x9=36	9x4=36	36÷9=4	36÷4=9
Style	Style	Style	Style
Rummy	Rummy	Rummy	Rummy
4x3=12	3x4=12	12÷3=4	12÷4=3
Style	Style	Style	Style

Rummy Style [playing cards] (9 of 9)

Rummy	Rummy	Rummy	Rummy
3x2=6	2x3=6	6÷3=2	6÷2=3
Style	Style	Style	Style
Rummy	Rummy	Rummy	Rummy
4x2=8	2x4=8	8÷2=4	8÷4=2
Style	Style	Style	Style

Appendix A

Manipulatives Used in Activities

This section is designed to define the manipulatives that are utilized throughout the activities in this book. Resources are also given for where they may be purchased if you choose not to make them. Page numbers are also included that indicate where the manipulative is used.

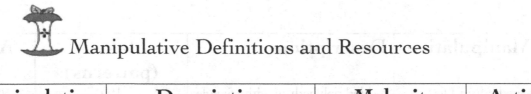

Manipulative Definitions and Resources

Manipulative	Description	Make it (patterns)	Activity Page
Base Ten Blocks	These 3-dimensional wooden or plastic blocks come in units (1's), longs (10's), flats (100's) and blocks (1,000).	None	41,135
Colored Squares (1")	1" colored (6 colors) squares usually made out of plastic.	None	11
Counters	Buttons, beans, washers, beads, small blocks – anything that could be easily used as counters.	None but can be found in hardware, craft, grocery, and teacher supply stores	3.23.25.27.58.95. 99.107.127
Counters – two sided	Plastic round discs that are yellow on one side and red on the other.	None	90.190
Die	6-sided cube usually with dots representing the numbers 1-6	67	65.127.190
Double- Nine Dominoes	This domino set consists of 55 dominoes with the numbers represented on them ranging from 0(blank) to a double 9.	49-50	48
Game Markers	Sometimes called 1"pawns used to mark spaces in games. Counters can also be used as long as they are different to distinguish each player.	None	65.140.177
Graph Paper	Special paper that is lined with measured squares. Different size squares serve different purposes.	13.14.75	10.11.23.25.27.29 .73.77.80.86.103. 166.170.183.190. 195.208
Interlocking Cubes	1 cm 3-dimensional cubes that can connect from all sides. They usually come in four colors.	None	7.47
Lego Blocks	Special blocks that typically have 6 studs that interlock with another block. They come in different colors, sizes and widths.	None	10
Lego Building Plates	Flat bases with studs that legos can snap into. They come in different sizes and colors.	None	10

Manipulative	Description	Make it (patterns)	Activity Page
Octahedron Die	8 sided polyhedron with triangle shaped sides and can be used for dice.	100	99.181.183
Play Money	Paper money that represents the different denominations of real currency.	184-187	183
Popscicle Sticks	Flat, wooden sticks that are 4-1/2 x 3/8 in dimension.	None	150
Tiles	1' square tiles that can be plastic or ceramic with math numbers and symbols	146-148	47.150
Unifix Cubes	These cubes interlock on one side and come in 10 different colors.	None	7.47

Resources for purchasing manipulatives

- Amazon – www.amazon.com
- Best-Deal.com - http://www.best-deal.com
- Carson-Dellosa - http://www.carsondellosa.com
- Classroom Direct - https://store.schoolspecialty.com
- Didax- www.didax.com
- Discount School Supply – www.discountschoolsupply.com
- EAL Education – www.ealeducation.com
- ETA Hand2Mind - http://www.hand2mind.com/
- Nasco - http://www.enasco.com/math/
- Learning Resources.com - http://www.learningresources.com/category/subject/math/manipulatives.do

Appendix B

Children's Literature to Develop and Reinforce Multiplication and Division

Children's Literature to Develop and Reinforce Multiplication

Anno, M. (1999*). Anno's Mysterious Multiplying Jar*. New York, NY: Penguin Putnam Books for Young Readers.
A powerful picture book that starts with a jar and teaches the concept of multiplication and factorials.

Appelt, K. (1998). *Bats On Parade*. New York, NY: Morrow Junior Books.
The Marching Bat Band performs on a midsummer's night creating multiplication formations from two times to ten times tables.

Brenner, M. (2000). *Stacks of Trouble*. New York, NY: Kane Press.
When the dishwasher is broken, Mike learns dirty dishes can multiply. The chore he agreed upon turns out to be a lot more work than he thought.

Calvert, P. (2011). *The Multiplying Menace*. Watertown, MA: Charlesbridge.
With mischief from Rumpelstiltskin and his multiplying stick creating multiplying problems, Peter and his dog zero try and save the kingdom.

Chae, I.S. (2003). *How do you Count a Dozen Ducklings*. Morton Grove, IL: Albert Whitman.
A mama duck has too many ducklings to count so she learns new ways to count by twos, threes, fours and sixes.

Dodds, D. (2004). *Minnie's Diner: A Multiplying Menu*. Cambridge, MA: Candlewick Press.
Down on a farm, where chores were to be done. The concept of doubling is taught through rhyme and humor with Papa McFay's son's appetites.

Esham, B. (2008). *The Adventures of Everyday Geniuses Last to Finish: A Story about the Smartest Boy in Math Class*. Perry Hall, MD: Mainstream Connections Publishing.
Max has a difficult time with his math-timed tests but to everyone's surprise he has been solving algebra problems in his spare time. A delightful story about name potential.

Giganti, P. (1994). *Each Orange had Eight Slices*. New York, NY: Mulberry Big Books.
Filled with fun and colorful multiplication problems for readers to solve.

Horton, J. (2009). *Math Attack*. Canada: Douglas & McIntyre, LTD.
A comical tale of a young girl exploring numbers that add humor to learning the multiplication table.

Kampelien, T. (2006). *I Can Multiply, it's not a Lie*. Abdo Group.
Colorful examples of a variety of multiplication problems to build basic understanding.

Matthews, L. (1978). *Bunches and Bunches of Bunnies*. New York, NY: Dodd Mead.
Bunnies marching in bunny lines invite the reader to count and learn the basic
multiplication operations 1 to 12.

McElligott, M. (2005). *The Lion's Share: A tale of Halving Cake and Eating it*. New
York, NY: Walker Publishing Co.
Ant receives an invitation from Lion for dinner but Ant barely gets a crumb. A
tale of halving and doubling.

Mills, C. (2002). *7x9= Trouble*. New York, NY: Farrar Straus Giroux.
A chapter book about Wilson, a third grader and his adventures with learning
his times tables and trying to pass his timed-multiplication tests.

Murphy, S. (1996). *Too Many Kangaroo Things to do*. New York, NY: Harper
Collins.
A birthday party for kangaroo provides the reader with fun things to multiply as
everyone as too many things to do.

Neuschwander, C. (1998). *Amanda Bean's Amazing Dream*. New York, NY:
Scholastic Press.
Amanda Bean likes to count everything until she has a crazy dream that makes
her realize multiplication is a much faster way to count. The book contains math
activities in the back for parents and teachers by Marilyn Burns.

Pallotta, J. (2002). *The Hersey's Milk Chocolate Multiplication book*. New York, NY:
Scholastic.
The Basics of multiplication are taught with a Hershey's Milk chocolate bar.

Park, M. (2007) *Now for my Next Number*. Salt Lake City: Utah. Great River Books.
A collection of songs and rhymes to learn multiplication.

Pittman, H.C. (1986). *A Grain of Rice*. New York, NY: Bantam Doubleday Dell
Books for Young Readers.
A clever Pong Lo multiples a grain of rice into a fortune which baffles an
Emperor. A tale of doubling.

Rawlingston, M. (1998). The Numberland Tales: A collection of Stories about
Learning Multiplication. [Kindle] Paradox Theatre.
The Numberland Tales is a series of stories featuring a boy called Danny
Starbright who has difficulty with multiplication. He has some exciting, and
sometimes amusing adventures, during the course of which, he learns how
multiplication works, and so finally understands and learns the times tables. Ice
Cream and Spiders: Two Times Tables/ Three Barts and a Bogie: Three Times
Tables/ Four Twits a Dragon and a Princess: Four Times Tables/ Cakes and
Butterflies: Five Times Tables/ The Tentown Adventures: Ten Times Tables.

Schwartz, R. (2010). *You Can Count on Monsters: The First 100 Numbers and their
Characters*. New York, NY: CRC Press.

Colorful and playful monsters create a visual way to explore factoring, prime and composite numbers and the basic multiplication.

Slade, S. (2011). *Multiply on the Fly*. Sylvian Dell Publishing.
From Spittlebugs to Luna moths, multiplication is explored through the world of insects.

Stamper, J. (2003). *Breakfast at Danny's Diner*. New York, NY: Grosset & Dunlap.
Danny's twin niece and nephew help him prepare to open his diner one morning but are soon overwhelmed by the work, until they put their multiplication skills to work.

Tang, G. (2002). *The Best of Times*. New York, NY: Scholastic Press.
Clever and strategic rhymes teach how to multiply zero through ten without memorizing the facts. .

Trivett, J. (1975). *Building Tables on Tables*. New York, NY: Crowell.
Multiplication tables are transformed into easy and fun games to bring the basic concepts of multiplication alive.

Children's Literature to Develop and Reinforce Division

Calvert, P. (2011). *The Multiplying Menace Divides*. Watertown, MA: Charlesbridge.
This enchanted fairy tale, full of division by whole numbers and fractions takes Jack and his dog Zero on an adventure of trying to save the kingdom.

Cato, S. (1999). *Division*. Minneapolis, MN: Carolrhoda Books.
Mia and her dog Popcorn encounter a series of adventures that teach them about division with the help of her number friend digit.

Dodds, D.A. (2005). *The Great Divide*. Cambridge, MA: Candlewick Press.
A whimsical tale of a cross-country race of 40 participants that encounter various obstacles that divide them until one crosses the finish line.

Froman, R. (1978). *The Greatest Guessing Game*. New York, NY: Crowell.
Students are invited to guess, revise, and check their answers while exploring a world of division fun and division scenarios.

Hutchins, P. (2006). *The Doorbell Rang*. New York, NY: Greenwillow Books.
Sharing Ma's delicious cookies each time the doorbell rings introduces young to division problems filled with friend and sharing until they better not open the door anymore, but Grandma shows up with more to continue the adventure.

Jo, N. & Tchen, R. (2001). *How Hungry are You?*. New York, NY: Scholastic, Inc.
A picnic with two friends turns into a party with lots of friends and problems dividing up the irresistible snacks.

Kompelien, T. (2007). *I Can Divide, I need no Guide!* Edina, MN: ABDO Publishing Company.
The basics of division are illustrated through sharing a variety of objects from peas to eggs.

McElligott, M. (2007). *Bean Thirteen*. New York, NY: G.P. Putnam's Sons.
A delightful story to introduce division as two bugs, Ralph and Flora, try to divide thirteen beans evenly. They keep inviting more to create more division problems.

Murphy, S. (1997). *Divide and Ride*. New York, NY: Harper Collins Publishing.
Predivision skills are taught as Carnival fun teaches 11 friends how to divide with a 2-seat roller coaster and a 4-seat teacup ride.

Murphy, S. (1999). *Jump, Kangaroo, Jump*. New York, NY: Harper Collins Publishers.
Kangaroo has a great time at camp with is group of Australian friends as they divide into lots of different teams for a variety of fun games. Additional basic division activities also provided.

Nagda, A. (2007). *Cheetah Math: Learning about Division from Baby Cheetahs*. New York, NY: Henry Holt and Co.
Using the story of two cheetahs coming to the nursery at the San Diego Zoo, this book illustrates introduces a variety of division concepts used in the care of the cubs.

Pinczes, E. (1995). *A Remainder of One*. Boston, MA: Houghton Mifflin.
A tale of a queen insisting her army marches in even lines and private Joe keeps trying different formations so he is not left out of the parade.

Rocklin, J. (1997). *One Hungry Cat*. New York, NY: Scholastic.
Tom the cat bakes all sorts of snacks to try and share with two friends. When he gobbles up some before, he is faced with new division problems. Includes division activities at the back of the book by Marilyn Burns.

Wingard-Nelson, R. (2005). *Division Made Easy*. Berkeley Heights, NJ: Enslow Elementary.
A variety of division concepts and problems are presented with colorful illustrations and graphics